Topics in Boundary Element Research

Edited by C.A. Brebbia

Volume 4
Applications in Geomechanics

With 87 Figures

Springer-Verlag Berlin Heidelberg New York
London Paris Tokyo

Editor:
Dr. Carlos A. Brebbia

Computational Mechanics Institute
Wessex Institute of Technology
Ashurst Lodge
Ashurst,
Hampshire SO4 2AA
England

ISBN 13 : 978-3-642-83014-3 e-ISBN-13 : 978-3-642-83012-9
DOI : 10.1007 / 978-3-642-83012-9

Library of Congress Cataloging-in-Publication Data
(Revised for vol. 4)
Topics in boundary element research.

Includes bibliographies and indexes.
Contents: v. 1. Basic principles and applications – v. 2. Time-dependent and vibration
problems – [etc.] – v. 4. Applications in geomechanics
1. Boundary value problems. 2. Transients (Dynamics) 3. Vibration. I. Brebbia, C. A.
TA347.B69T67   1984   620'.001'51535   84-10644
ISBN 0-387-13097-7 (v. 1.: New York)

© Springer-Verlag Berlin, Heidelberg 1987
Softcover reprint of the hardcover 1st edition 1987

Typesetting: Acso Trade Typesetting Ltd., Hong Kong

2161/3020-5 4 3 2 1 0

Contributors

R. Abascal	Universidad de Sevilla, Spain	(Chap. 2)
M.C. Au	Carleton University, Canada	(Chap. 5)
D.E. Beskos	University of Patras, Greece	(Chap. 1)
A.H.-D. Cheng	University of Delaware, USA	(Chap. 6)
J. Dominguez	Universidad de Sevilla, Spain	(Chap. 2)
D.L. Karabalis	University of South Carolina, USA	(Chap. 1)
M. Kemblowski	Shell Development Company, Westhollow Research Center, Houston, USA	(Chap. 4)
G.P. Lennon	Lehigh University, USA	(Chap. 8)
K. Mizumura	Kanazawa Institute of Technology, Japan	(Chap. 7)
N. Nishimura	Kyoto University, Japan	(Chap. 3)
A.P.S. Selvadurai	Carleton University, Canada	(Chap. 5)

Preface

The first volume of this series dealt with the Basic Principles of Boundary Elements, while the second concentrated on time dependent problems and Volume three on the Computational Aspects of the method. This volume studies the applications of the method to a wide variety of geomechanics problems, most of which are ideally suited for boundary elements demonstrating the potentiality of the technique.

Chapter 1 deals with the application of BEM to three dimensional elastodynamics soil-structure interaction problems. It presents detailed formulations for rigid, massless foundations of arbitrary shape both in the frequency and time domains. The foundations are assumed to be resting on a linearly elastic, homogeneous, isotropic half-space and be subjected to externally applied loads on obliquely incident body. The chapter reviews the major advances in soil foundation interaction presents a series of numerical results and stresses the practical application of BEM, pointing out the high accuracy and efficiency of the technique, even when using coarse mesh discretizations.

Chapter 2 presents a number of numerical results after a brief review of the basic equations governing the dynamics formulation of BEM. The general foundation stiffness problem is presented and stiffness coefficients are calculated for 3D, 2D and axisymmetric foundations. Then the response of foundations to travelling harmonic waves is studied, including the cases of embedded, strip and rectangular foundation. Another section reviews the use of time discretization BEM for the analysis of foundations. This chapter presents the state of the art on the topic of soil foundation interaction including diagrams obtained using numerical results and of utility in engineering practice.

Chapter 3 reviews the BEM formulation for consolidation problems, including Biot's method and the more practical approaches used by soil mechanicists. The formulations are described in detail and some numerical examples are presented. The merit of the chapter lies not only in the up to date presentation of all available theory but in the attempt by the author to relate together all the different approaches.

A review of the Boundary Element Model for salt water intrusion is attempted in the following chapter. This is a particularly attractive application for BEM as it involves moving boundaries. BEM models are concluded to cover a broad spectrum of salt water – intrusion problems and to be superior to finite difference or finite element models for this type of interface problem.

Chapter 5 studies a problem of interest in soil mechanics and others, i.e. the application of the BEM to the study of non-linear interface behaviour between two material regions. The non-linear interface response is modelled either by Coulomb frictional behaviour or by interface plasticity. The results presented by the authors illustrate the manner in which the non-linear phenomena at the interface contributes to the global non-linear response in the composite.

The last three chapters deal with flow through porous media. Chapter 6 presents some innovative ideas to deal with heterogeneities in these flows which is a problem for which BEM was thought to be inappropriate. The approach is based on using simple transformations and the author presents a table listing some of them. Notice that the method can be used for multizoned regions as well.

The effects of spatial distribution of soil infiltration properties and rainfall rates on the performance of a catchment area are studied in Chap. 7. Several important conclusions are obtained using recharge sources varying in time and space.

The last chapter, 8, revises the BEM formulation of unconfined groundwater flow, involving a moving, non-linear free surface as well as using the vertically integrated approach.

The chapters in this volume bring together a series of recent advances made in the applications of BEM for soil mechanics, consolidation, foundations and groundwater flow problems. Research in these problems has progressed to a degree such that the BEM can be employed as an engineering analysis tool.

Southampton, June 1987   **Carlos A. Brebbia**
                                                                        Editor

Contents

Chapter 1

Three-Dimensional Soil-Structure Interaction by Boundary Element Methods

by D.L. Karabalis and D.E. Beskos

Abstract

The application of the Boundary Element Method to the linear three-dimensional soil-structure interaction problem is discussed. Detailed formulations for rigid, surface, massless foundations of arbitrary shape are given in both frequency and time domains. In both cases the foundations are assumed to rest on a linear elastic, homogeneous, and isotropic half-space and are subjected to either externally applied loads or obliquely incident body or surface waves. Results obtained by the above approaches as well as by other well established techniques are given in a comparison study. More general problems involving massive foundations and superstructures are also presented in the general framework of a substructure formulation.

1.1 Introduction

During the fifty years that have passed since Reissner's pioneering work [1, 2], a significant amount of research activity has been devoted to the field of dynamic soil-structure interaction (SSI). As a result of that, a number of techniques are available today for solving a variety of problems within the general area of SSI. These techniques range from the simplified to the very sophisticated and are even available commercially. However, due to the highly complex character of the real problem and our limited knowledge of the several factors which might affect its solution, all of the existing models deal with highly idealized portions of it.

In very general terms, the dynamic SSI problem consists of computing the stresses or the displacements that would occur at the contact surface between the foundation and the soil when the soil-structure system is subjected to general transient external loads or obliquely incident seismic waves. These contact stresses do not only affect the motion of the overlying structure but deform the neighboring soil as well. This soil deformability has been proven to influence substantially the overall behavior of the soil-structure system and solely accounts for the SSI phenomenon. There are cases, of course, in which the deformations of the soil are insignificant compared to the distortions within the structure itself or to the uncertainties involved in specifying these deformations, and thus one can assume that the structure is supported on a rigid soil. There exists evidence in several cases, however, such as in tall buildings supported on moderately soft soils, extensive mat foundations, or adjacent multiple footings, to name but a few, that SSI phenomena occur and are of importance. In order to account for these phenomena in some sort of a "realistic" way, a wide variety of factors should be taken into consideration. For example, with regard to the geometry, the three-dimensional character of the problem, and the arbitrary shape and arrangement of adjacent structures, are only a few of the factors which should be considered. Further, a mechanical and material characterization of the soil-structure system should take into consideration rather well-known properties of the related structural elements and soil deposits, such as concrete slabs and domes, steel frames, soil stratification, etc., as well as soil properties not yet thoroughly understood, such as nonlinear dynamic stress-strain relationships and dynamic consolidation. Discontinuities of stresses as a result of lateral separation or uplift at the contact surface between the soil and the foundation should also be considered.

This article addresses the problem of the interaction between a three-dimensional rigid, surface foundation and a homogeneous, isotropic, linear elastic half-space, where no uplift or lateral separation can occur at the contact surface between the two media. An almost complete account of the research done in this area is given in the book of Richart, Hall and Woods [3], which reviews the work up to 1968, as well as in several other updated references, e.g. [4–8, 80]. In general it has been customary to classify the SSI methods into three major groups: the "continuous" models, the "half-space approximate" models, and the "discrete finite" models.

In the first group, the "continuous" models, the dynamic stiffnesses or flexibilities of the foundation are obtained in a complex form by solving semi-analytically the governing equations of the complete mixed boundary value problem. Collins [9], Robertson [10], Gladwell [11], and others, used the above approach in order to obtain solutions for the various modes of vibration of a rigid circular disc on an elastic half-space. Veletsos and Verbic [12] also considered the problem of the rigid circular footing but on a viscoelastic half-space, thus introducing the effect of the soil material damping in addition to the radiation or geometrical damping. The solutions reported in Refs. [9–12] were obtained under the assumption of relaxed boundary conditions, i.e., the various modes of oscillation of the circular disc on the half-space are assumed to be independent from each other. The coupling effect between the tangential and rocking oscillations of a rigid circular disc was studied by Veletsos and Wei [13] and Luco and Westmann [14], while Luco and Westmann [15] considered the problem of complete bond between the soil and a strip foundation. The "continuous" models, although successful in presenting convenient semi-analytical solutions, are of small practical importance since they can only be applied to relatively simple cases involving harmonic motions, simple boundaries, and linear elastic or viscoelastic materials.

In the second general group of approaches, the "half-space approximate" models, the problem of rigid foundations of arbitrary shape is introduced. The various existing techniques that fall into this major category proceed first by discretizing the contact surface between the soil and the foundation into a number of "subregions" of simple geometrical form. This step is necessary in order to replace the unknown contact stresses under each "subregion" with an approximate distribution of stresses, usually of constant variation or with only a concentrated force at the center of the "subregion", and at the same time to make possible the application of analytical solutions, e.g., from Lamb's problem [16], to express the stresses and displacements under each "subregion". Finally, appropriate compatibility and equilibrium conditions are applied among all the "subregions" in order to simulate the rigid body motion of the foundation. This basic methodology was introduced as early as 1965 by Lysmer [17], who studied the vertical flexibility of a circular rigid disc on an elastic half-space by considering several uniformly loaded concentric rings. Numerous other applications of the method have been reported for various problems in the general area of SSI. For example, the works of Elorduy, Nieto and Szekely [18], Wong and Luco [19], Gaul [20], Kitamura and Sakurai [21, 22], Adeli et. al. [23], and Hamidzadeh-Eraghi and Grootenhuis [24], are among the many analyzing the problem of a single surface foundation of arbitrary shape under externally applied forces. Cross-interaction effects between adjacent foundations were also studied by Savidis and Richter [25] and Gantayat and Kamil [26], among others. Wong and Luco [27, 28] considered the effect of oblique seismic waves on surface, rigid, rectangular foundations, and Bielak and Coronato [29] the motion of two rectangular foundations on a viscoelastic half-space under the influence of oblique SH and Rayleigh waves. The complete problem of an entire structure, supported on a set of rigid surface foundations, and subjected to nonvertically incident seismic waves was also formulated and solved by Werner, Lee, Wong and Trifunac [30] and Luco and Wong [31].

The third major group of methods applied to SSI, the "discrete finite" models, consists of the Finite Element Method (FEM) and the Finite Difference Method (FDM), the two most widely used numerical methods for general wave propagation problems, with the former being the most popular of the two. From the many reported applications of the FEM to general SSI problems, very few have dealt with the three-dimensional rigid surface foundation case, e.g., Roesset and Gonzalez [32], Gupta, Penzien, Lin and Yeh [33], and Dasgupta and Rao [34]. An SSI analysis using the FEM appears, in principle, to be very effective because of the distinct advantages that this method presents in handling complex foundation geometries and soil inhomogeneities. In addition, a time domain FEM analysis, through a step-by-step integration, provides the means for studying nonlinear SSI problems. However, the use of the FEM in SSI presents serious drawbacks due to the fact that the semi-infinite soil medium is represented by a finite size model and thus wave reflecting boundaries are artificially introduced. This inherent deficiency can be compensated for in part by the use of very large models, or non-reflecting (transmitting) boundaries [35, 36], or special infinite elements [37, 38]. None of these improvements succeeds in completely eliminating the problem, and in fact they complicate the application of the FEM and make it highly uneconomical. Hybrid finite element models for three-dimensional SSI have also been reported, e.g., [33, 39–45]. Detailed comparisons of the two basic methods for solving SSI problems, namely, the "half-space approximate" method and the FEM, have been reported by Hadjian, Luco and Tsai [46], favoring the former method, and by Seed, Lysmer and Hwang [47] favoring the latter.

Several applications of the FDM in the general area of SSI have also been reported [48–52]. This method, like the FEM, presents us with the built-in problem of the artificially introduced wave reflecting boundaries, and attempts have been made to correct it by the use of specially developed two-dimensional nonreflecting boundaries [51–52]. However, due to its inferior handling of the complicated geometries frequently encountered in SSI problems, the popularity of the FDM in this area remains limited.

In this brief literature review and classification of methods currently applied to the SSI analysis, the discussion of the Boundary Element Method (BEM) was left for last since it constitutes the main subject of this article. The BEM is the latest addition to the group of major numerical methods used for dynamic SSI analyses, and appears to be ideally suited for modelling the half-space soil medium, particularly in three-dimensional problems. The reason, of course, is that this method requires discretization of only the boundary of the domain of interest, which in the present case is usually the contact surface between the soil and the foundation, and takes into account automatically the radiation condition due to the use of the fundamental singular solutions (Green's functions). Further, the immediate consequences of requiring a discretization only at the surface are, first, that the dimensions of the problem are reduced by one and, second, a minimum amount of discretization is required. These are distinct advantages over "domain" type methods, such as the FEM and the FDM, which require a discretization of both the interior of the domain of interest and its surface, and artificial nonreflecting boundaries.

To the writers' best knowledge, Dominguez [53, 54] was the first to apply the BEM in order to compute, in a frequency domain formulation, the impedances of two- and three-dimensional surface and embedded, rigid, rectangular foundations resting on a uniform linear elastic half-space. Ottenstreuer and Schmid [55] and Ottenstreuer [56], following the same approach, studied, respectively, the problems of rigid, surface foundations and of cross-interaction between two rigid, surface, rectangular foundations. Apsel [57] obtained the dynamic stiffnesses of rigid cylindrical foundations embedded in a uniform or layered viscoelastic half-space using an indirect BEM. Both relaxed and non-relaxed boundary conditions were investigated in Refs. [53–57, 79]. Recently Wolf and Darbre [58, 59] studied the application of three BEM formulations, i.e., the weighted-residual technique and the direct and indirect BEM, to the problem of a two-dimensional foundation embedded in a layered elastic half-plane. A time domain BEM for three-dimensional SSI analysis was first reported by Karabalis and Beskos [60]. This method, which uses step-by-step integration in time, has also been applied to the interaction problems of a linear elastic half-space with rigid surface [60], rigid embedded [5, 61], and flexible surface foundations [62].

Hybrid methods involving the FEM and the BEM have also been reported, e.g., [63, 64]. In the SSI case, the BEM is used to describe the unbounded exterior domain and the FEM the bounded interior one, thus taking full advantage of the merits of both methods. This idea, which is similar to the hybrid methods of Refs. [33, 39–45], has been utilized by various workers in the field such as Varadarajan and Singh [65], Mita and Takanashi [66], and Gaitanaros and

Karabalis [78], to name but a few in the frequency domain, and Karabalis and Beskos [62] in the time domain.

It should be emphasized at this point that, in contrast to all the other frequency domain methods, the general time domain methodology reported in Refs. [5, 60–62] presents a natural and direct way of dealing with transient problems and, in addition, can form the basis for an extension to nonlinear SSI analysis where nonlinearities could occur in the soil and/or the structure. Of course a nonlinear SSI analysis can be done, in principle, by a direct application of the time domain FEM. However, due to its deficiencies in simulating unbounded regions, the use of this method in a true three-dimensional SSI analysis is associated with prohibitive high costs and frequently with inaccuracies. The time domain approach of Veletsos and Verbic [67], which is restricted to circular foundations only, is based on impulse response functions in a convolution formulation which precludes its extension to nonlinear soil problems.

In the following sections the problem of the dynamic response of a three-dimensional rigid, surface, massless foundation on a linear elastic, homogeneous and isotropic half-space due to both externally applied loads and obliquely incident seismic waves, is formulated using the BEM. Both the frequency and the time domain approaches are presented in detail following the developments reported in Refs. [53, 54] and [5, 60], respectively. An extension of the massless foundation problem to more general problems involving massive foundations and/or superstructures is presented on the basis of a soil-structure interaction superposition principle. Finally, comparisons between the time domain and frequency domain BEM as well as other well known methodologies are given and conclusions on their accuracy and efficiency are drawn.

1.2 Field Equations

In this section the classical integral formulation of the elastodynamic problem, in both the time and the Fourier transformed or frequency domain, is reviewed. The following developments are based on the book by Eringen and Suhubi [68], for the time domain, and the work of Cruse and Rizzo [69], for the frequency domain. The standard index notation has been adopted where summation is assumed over repeated indices, commas indicate spatial differentiation, dots indicate differentiation with respect to time t, Greek subscripts take the values 1 and 2, and Latin subscripts take the values 1 to 3.

The governing equations, in terms of displacements, for a homogeneous, isotropic, linear elastic solid occupying a regular region R with a surface B can be written in the form

$$(c_1^2 - c_2^2)u_{i,ij} + c_2^2 u_{j,ii} + f_j = \ddot{u}_j, \tag{1}$$

where f_j is the body force vector per unit mass, the dilatational and shear wave velocities are given, respectively, as

$$c_1^2 = (\lambda + 2\mu)/\rho, \quad c_2^2 = \mu/\rho, \tag{2}$$

with μ and λ being the Lame constants, ρ being the mass density, and

$$u_i = u_i(x, t), \tag{3}$$

where x denotes the position vector. The stresses and the displacements should satisfy the following boundary conditions:

$$\begin{aligned} t_{(n)i}(x, t) &= p_i(x, t), \quad x \in B_t \\ u_i(x, t) &= q_i(x, t), \quad x \in B_u, \end{aligned} \tag{4}$$

where t_{ij} is the stress tensor, n is the normal vector on a differential element of the surface B, and $B_t + B_u = B$; and initial conditions:

$$\begin{aligned} u_i(x, 0^+) &= u_{0i}(x), \quad x \in R \cup B \\ \dot{u}_i(x, 0^+) &= v_{0i}(x), \quad x \in R \cup B. \end{aligned} \tag{5}$$

The constitutive equation for the above solid reads

$$t_{ij} = \rho(c_1^2 - 2c_2^2)u_{m,m}\delta_{ij} + \rho c_2^2(u_{i,j} + u_{j,i}), \tag{6}$$

where δ_{ij} is the Kronecker delta, and the stress vector is given by

$$t_{(n)i} = t_{ij}n_j. \tag{7}$$

1.2.1 Time Domain Integral Representation

In order to establish an integral representation solution of Eq. (1), it is first necessary to specify the required fundamental solution that will be used. In this formulation, the fundamental singular solution of Eq. (1) in an infinite solid medium due to a concentrated body force will serve in this capacity. Such a body force can be expressed as

$$\rho f(x, t) = f(t)\delta(x - \xi)e, \tag{8}$$

where x and ξ are points in a body of infinite extent, δ is the Dirac delta function, e is the direction in which the above force is applied, and $f(t)$ is its time variation. Application of Eq. (8) as a body force in Eq. (1) will yield the response of the infinite medium in the form

$$u_i = u_{ij}^0 e_j, \tag{9}$$

where the second order displacement tensor, usually called the fundamental singular solution of the elastodynamic equations or Stokes' displacement tensor, is given by

$$\begin{aligned} u_{ij}^0(x, t; \xi | f) = \frac{1}{4\pi\rho} &\left\{ \left(\frac{3r_i r_j}{r^3} - \frac{\delta_{ij}}{r} \right) \int_{c_1^{-1}}^{c_2^{-1}} \lambda f(t - \lambda r) \, d\lambda \right. \\ &\left. + \frac{r_i r_j}{r^3} \left[\frac{1}{c_1^2} f\left(t - \frac{r}{c_1} \right) - \frac{1}{c_2^2} f\left(t - \frac{r}{c_2} \right) \right] + \frac{\delta_{ij}}{r c_2^2} f\left(t - \frac{r}{c_2} \right) \right\}, \end{aligned} \tag{10}$$

where

$$r_i = x_i - \xi_i, \quad r^2 = (x_i - \xi_i)(x_i - \xi_i). \tag{11}$$

The Stokes' displacement tensor expresses the i-component of the displacement that occurs at point x and time t due to a concentrated force of magnitude $f(t)$ applied at point ξ in the j-direction. The function $f(t - s)$, as used in Eq. (10), is assumed to be time retarded, i.e., it is non-zero only if $t - s > 0$. The Stokes' stress tensor related to u_{ij} can be obtained through substitution of Eq. (10) into the constitutive Eq. (6). Following the above notation the Stokes' stress tensor can also be written in the form

$$t_{ijk}^0 = t_{ijk}^0(x, t; \xi | f). \tag{12}$$

The pair of fundamental singular solutions $[u_{ij}^0, t_{ijk}^0]$ possesses the properties of causality, and translation, and is called the Stokes' state of quiescent past [68].

Using Stokes' state of quiescent past as one of the two distinct elastodynamic states in Betti's reciprocal theorem, the other one being the actual state, one can derive a solution of the problem posed by Eqs. (1), (4), and (5) in the form of Love's integral representation

$$\varepsilon(\xi) u_k(\xi, t) = \int_B \{ u_{ik}^0[x, t, \xi | t_{(n)i}(x, t)] - t_{(n)ik}^0[x, t; \xi | u_i(x, t)] \} \, dB(x)$$
$$+ \rho \int_R u_{ik}^0[x, t; \xi | f_i(x, t)] \, dR(x)$$
$$+ \rho \int_R [v_{0i}(x) U_{ik}(x, t; \xi) + u_{0i}(x) \dot{U}_{ik}(x, t; \xi)] \, dR(x), \tag{13}$$

where

$$\varepsilon(\xi) = \begin{cases} 1 & \text{if } \xi \in R \\ \frac{1}{2} & \text{if } \xi \in B \\ 0 & \text{if } \xi \in R \cup B, \end{cases} \tag{14}$$

$$t_{(n)ik}^0 = t_{ijk}^0 n_j, \tag{15}$$

is the traction tensor of the Stokes' state. The tensor U_{ik}, also called Green's tensor, can be obtained as a special case of Stokes' displacement tensor by setting $f(t) = \delta(t)$ in Eq. (10).

For problems in SSI analysis it is usually assumed that both the body forces and the initial conditions are zero, and thus Eq. (13), written for points ξ on the boundary, is simplified to

$$\tfrac{1}{2} u_k(\xi, t) = \int_B \{ u_{ik}^0[x, t; \xi | t_{(n)i}(x, t)] - t_{(n)ik}^0[x, t; \xi | u_i(x, t)] \} \, dB(x). \tag{16}$$

Equation (16) relates boundary displacements and tractions and thus it constitutes the appropriate boundary integral equation (BIE) to be used in a time domain formulation of the foundation problems under consideration.

1.2.2 Frequency Domain Integral Representation

The Fourier transform of a function $f(x, t)$ is defined as

$$\bar{f}(x, w) = F\{f(x, t)\} = \int_{-\infty}^{+\infty} f(x, t) e^{-iwt} \, dt, \tag{17}$$

where w is the Fourier transform parameter, i.e., the frequency. Using this definition the general elastodynamic problem stated by Eqs. (1), (4), and (5) can be transformed into a static-like problem with respect to the frequency parameter w. Thus, assuming quiescent initial conditions as in the previous section, i.e., $u_{0i}(x) = v_{0i}(x) = 0$, the governing equations are transformed into

$$(c_1^2 - c_2^2)\bar{u}_{i,ij} + c_2^2 \bar{u}_{j,ii} + \bar{f}_j = -w^2 \bar{u}_j, \tag{18}$$

and similarly the boundary conditions, Eq. (4), and the constitutive relations, Eq. (6), become, respectively,

$$\bar{t}_{(n)i}(x, w) = \bar{t}_{ij} n_j = \bar{p}_i(x, w), \quad x \in B_t$$
$$\bar{u}_i(x, w) = \bar{q}_i(x, w), \qquad x \in B_u, \tag{19}$$

$$\bar{t}_{ij} = \rho(c_1^2 - 2c_2^2)\bar{u}_{m,m}\delta_{ij} + \rho c_2^2(\bar{u}_{i,j} + \bar{u}_{j,i}). \tag{20}$$

Following the same basic steps as in the time domain case, one should specify the fundamental solutions that will be used before proceeding with the integral representation solution of Eq. (18). The fundamental solution of Eq. (18) for an infinite region subjected to a concentrated body force is again the natural choice. In this case the concentrated body force becomes the Fourier transform of Eq. (8), i.e.,

$$\rho \bar{f}_i(x, w) = \bar{f}(w)\delta(x - \xi)e_i, \tag{21}$$

and the fundamental solution of Eq. (18) can be expressed in the form

$$\bar{u}_i = \bar{u}_{ij}^* e_j. \tag{22}$$

The second order displacement tensor appearing in Eq. (22) is given by

$$\bar{u}_{ij}^*(x, \xi, w) = \frac{\bar{f}(w)}{4\pi\rho c_2^2}(\psi \delta_{ij} - \chi r_{,i} r_{,j}), \tag{23}$$

where

$$\chi = \left(-\frac{3c_2^2}{w^2 r^2} + \frac{3c_2}{iwr} + 1\right)\frac{e^{-iwr/c_2}}{r} - \left(\frac{c_2^2}{c_1^2}\right)\left(-\frac{3c_1^2}{w^2 r^2} + \frac{3c_1}{iwr} + 1\right)\frac{e^{-iwr/c_1}}{r} \tag{24}$$

$$\psi = \left(-\frac{c_2^2}{w^2 r^2} + \frac{c_2}{iwr} + 1\right)\frac{e^{-iwr/c_2}}{r} - \left(\frac{c_2^2}{c_1^2}\right)\left(-\frac{c_1^2}{w^2 r^2} + \frac{c_1}{iwr}\right)\frac{e^{-iwr/c_1}}{r}. \tag{25}$$

Insertion of Eq. (23) into Eq. (20) yields the traction companion tensor

$$\bar{t}_{ij}^*(x, \xi, w) = \frac{\bar{f}(w)}{4\pi}\left[\left(\frac{d\psi}{dr} - \frac{\chi}{r}\right)\left(\delta_{ij}\frac{\partial r}{\partial n} + r_{,j} n_i\right) - \frac{2}{r}\chi\left(n_j r_{,i} - 2r_{,i} r_{,j}\frac{\partial r}{\partial n}\right)\right.$$
$$\left. - 2\frac{d\chi}{dr}r_{,i} r_{,j}\frac{\partial r}{\partial n} + \left(\frac{c_1^2}{c_2^2} - 2\right)\left(\frac{d\psi}{dr} - \frac{d\chi}{dr} - \frac{2\chi}{r}\right)r_{,i} n_j\right]. \tag{26}$$

The fundamental solution pair $[\bar{u}_{ij}^*, \bar{t}_{ij}^*]$, as defined above, possesses the property of space translation.

By utilizing the transform domain form of Betti's reciprocal theorem, where one of the two required states is the $[\bar{u}_{ij}^*, \bar{t}_{ij}^*]$, the other one being the actual state, one can obtain a solution to the Fourier transformed elastodynamic problem in the form of the integral identity

$$\varepsilon(\xi)\bar{u}_j(\xi) = \int_B \bar{t}_{(n)i}(x)\bar{u}'_{ji}(x,\xi,w)\,dB(x) - \int_B \bar{u}_i(x)\bar{t}'_{ji}(x,\xi,w)\,dB(x)$$

$$- \int_R \rho\bar{f}_i(x,w)\bar{u}'_{ji}(x,\xi,w)\,dR(x), \tag{27}$$

where

$$\bar{u}'_{ij} = \bar{u}^*_{ij}/\bar{f}(w) \quad \text{and} \quad \bar{t}'_{ij} = \bar{t}^*_{ij}/\bar{f}(w). \tag{28}$$

Finally, on the assumption of zero body forces and points ξ on the boundary B, Eq. (27) is reduced to

$$\tfrac{1}{2}\bar{u}_j(\xi) = \int_B \bar{t}_{(n)i}(x)\bar{u}'_{ji}(x,\xi,w)\,dB(x) - \int_B \bar{u}_i(x)\bar{t}'_{ji}(x,\xi,w)\,dB(x) \tag{29}$$

which is the restraint equation between boundary tractions and displacements, i.e., the BIE in the Fourier transformed domain. Equation (29) provides the solution to a given dynamic problem in terms of the frequency parameter w. The time domain response can then be obtained through a Fourier synthesis of a sequence of frequency dependent solutions.

An analytic solution of the previously established boundary integral equations in the time, Eq. (16), or the frequency, Eq. (29), domains is almost impossible even for simple geometries and time variations of the related functions. The numerical treatment of Eqs. (16) and (29) is the subject of the next section.

1.3 Numerical Implementation

Toward a numerical solution of the boundary integral Eqs. (16) and (29), a spatial discretization is necessary in both the time and the frequency domain approaches. However, in contrast to the static-like frequency domain problem, the time domain formulation requires an additional discretization in time. The numerical procedures described in this section are following the developments reported by Karabalis [5], and Karabalis and Beskos [60] for the time domain, and Cruse [70] for the frequency domain case.

1.3.1 Time Domain Approach

The time domain BEM of Refs. [5, 60] consists of two basic steps: (i) a discretization of the real time axis into a sequence of equally spaced time intervals is applied, and further, the variation of the displacements and tractions over each time interval is assumed to be constant, and (ii) the boundary B of the domain of interest is discretized into a number of rectangular elements over each of which a constant distribution of displacements and tractions is considered. On the basis of these discretizations a time stepping solution of Eq. (16) can be established for the boundary displacements and tractions over each rectangular element and for each time step.

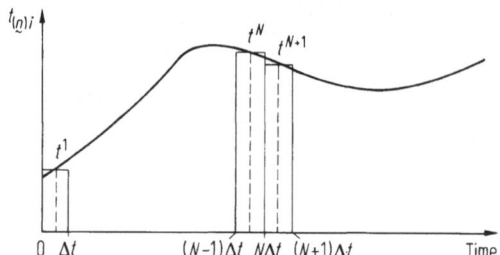

Fig. 1. Approximation of the surface traction $t_{(n)i}$ by a sequence of rectangular pulses (after Karabalis [5])

To illustrate the above outlined procedure, one might consider the time discretization of the continuous traction vector into a sequence of rectangular impulses as shown in Fig. 1. This approximation can also be expressed, for the time interval $(q-1)\Delta t < t < q\Delta t$, as

$$t_{(n)i}(x,t) \approx t_{(n)i}^q(x)\{H[t-(q-1)\Delta t] - H(t-q\Delta t)\}, \tag{30}$$

where

$$t_{(n)i}^q = t_{(n)i}[x,(q-0.5)\Delta t], \tag{31}$$

and represents the intensity of the rectangular impulse at time $t = (q-0.5)\Delta t$. Substituting the forcing function $f(t)$ in Eq. (10) by $t_{(n)i}(x,t)$ as it is expressed in Eq. (30), e.g., for $q = 1$, will yield the following time discretized expression for the Stokes' displacement tensor:

$$\begin{aligned}
u_{ik}^0(x,t;\xi|t_{(n)i}^1) = \frac{1}{4\pi\rho} &\left\{ \left(\frac{3r_ir_k}{r^5} - \frac{\delta_{ik}}{r^3}\right)\left[H\left(t-\frac{r}{c_1}\right)F_1 - H\left(t-\frac{r}{c_2}\right)F_2\right] \right. \\
&\cdot t_{(n)i}^1(\xi) + \frac{1}{c_1^2}\frac{r_ir_k}{r^3}H\left(t-\frac{r}{c_1}\right)t_{(n)i}\left(\xi,t-\frac{r}{c_1}\right) \\
&\left. + \frac{1}{c_2^2}\left(\frac{\delta_{ik}}{r} - \frac{r_ir_k}{r^3}\right)H\left(t-\frac{r}{c_2}\right)t_{(n)i}\left(\xi,t-\frac{r}{c_2}\right)\right\},
\end{aligned} \tag{32}$$

where

$$F_\beta = \begin{cases} [t^2 - (r/c_\beta)^2]/2, & \text{if } 0 < t-(r/c_\beta) \leqslant \Delta t \\ [2t(\Delta t) - (\Delta t)^2]/2, & \text{if } \Delta t < t-(r/c_\beta), \beta = 1,2. \end{cases} \tag{33}$$

A similar discretization in time of the boundary displacement vector $u_i(x,t)$, and direct substitution of it into the expression of the Stokes' stress tensor, indicated by Eq. (12), will result in a time discretized form of the Stokes' stress tensor [5].

The spatial discretization scheme utilized in 3-D SSI, for example, is shown in Fig. 2a, in which the contact surface between an arbitrarily shaped foundation and the half-space is discretized into an M number of rectangular elements.

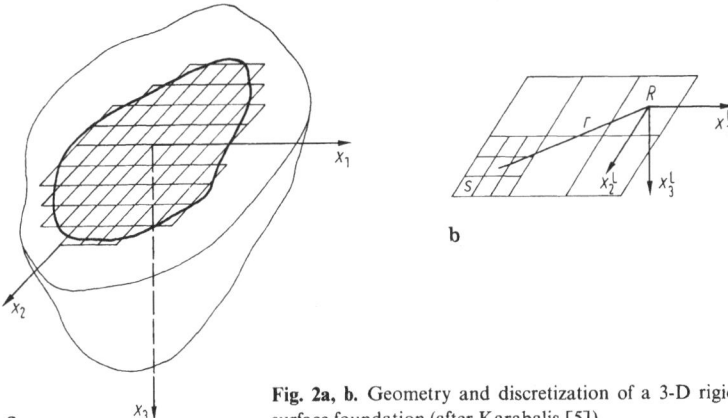

Fig. 2a, b. Geometry and discretization of a 3-D rigid surface foundation (after Karabalis [5])

On the assumption of constant variation of tractions and displacements over each rectangular element, and in view of Eq. (32), the boundary integral equation (16), when written for every boundary element R, will yield a system of M algebraic equations for each time step N, which can be best expressed in a matrix form as

$$\tfrac{1}{2}\{u^{N,R}\} = \sum_{n=q}^{N} \sum_{m=1}^{M} \{[G^{n-q+1,s}]\{t^{N-n+q,s}\} - [T^{n-q+1,s}]\{u^{N-n+q,s}\}\}$$

$$R = 1, 2, 3, \ldots, M, \tag{34}$$

where $\{u^{N,R}\}$ is the displacement vector at time step N, $\{t^{k,s}\}$ is the traction vector at time step k, and the superscripts R and s indicate, respectively, the "receiver" and "source" elements, as shown also in Fig. 2b. The pair of tensors $\{[G^{n,s}], [T^{n,s}]\}$ represents the time and space discretized equivalent of the Stokes' state of quiescent past. The displacement tensor $[G^{n,s}]$, for example, is given as

$$G_{ik}^{n,s} = \frac{1}{4\pi\rho}\{B_c^n\} \int_{(s)} \left(\frac{3r_i r_k}{r^5} - \frac{\delta_{ik}}{r^3}\right)\left[H\left(t - \frac{r}{c_1}\right)F_1 - H\left(t - \frac{r}{c_2}\right)F_2\right] d(s)$$

$$+ \frac{B_{c_1}^n}{c_1^2} \int_{(s)} \frac{r_i r_k}{r^3} d(s) + \frac{B_{c_2}^n}{c_2^2} \int_{(s)} \left(\frac{\delta_{ik}}{r} - \frac{r_i r_k}{r^3}\right) d(s), \tag{35}$$

where (s) is the area of the element s, and r is the distance from the center of the element R to each point of the element s, as shown in Fig. 2b. The coefficients B_c^n, $B_{c_1}^n$, and $B_{c_2}^n$ are used in order to specify the limits of spatial integration of the various quantities involved in Eq. (35) according to the kind of wave they represent, i.e.,

$$B_c^n, B_{c_\alpha}^n = \begin{cases} 1 & \text{if } r \in B_{n,c}, B_{n,c_\alpha} \\ 0 & \text{if } r \in B_{n,c}, B_{n,c_2}, \end{cases} \tag{36}$$

where

$$B_{n,c} = \{x; x \in B \text{ and } c_2(n-1) \cdot \Delta t < r < c_1 \cdot n \cdot \Delta t\}$$

$$B_{n,c_1} = \{x; x \in B \text{ and } c_1(n-1) \cdot \Delta t < r < c_1 \cdot n \cdot \Delta t\} \qquad (37)$$

$$B_{n,c_2} = \{x; x \in B \text{ and } c_2(n-1) \cdot \Delta t < r < c_2 \cdot n \cdot \Delta t\}.$$

Thus, the following 3×3 matrix is defined by Eq. (35):

$$[G^{n,s}] = \begin{bmatrix} G_{11}^{n,s} & G_{12}^{n,s} & G_{13}^{n,s} \\ G_{21}^{n,s} & G_{22}^{n,s} & G_{23}^{n,s} \\ G_{31}^{n,s} & G_{32}^{n,s} & G_{33}^{n,s} \end{bmatrix}, \qquad (38)$$

where its $G_{ik}^{n,s}$ term represents the i-component of the displacement vector which occurs at the center of the element R during the time step n, due to a unit rectangular impulse of the k-component of the traction vector acting over the element s during the first time step. For the case of surface foundations, i.e., $x_3 = 0$ as shown in Fig. 2, it is observed that the 9 terms of Eq. (38) reduce to 5 since

$$G_{13}^{n,s} \equiv G_{31}^{n,s} \equiv G_{23}^{n,s} \equiv G_{32}^{n,s} \equiv 0, \qquad (39)$$

and thus a decoupling of the vertical and horizontal motions occurs. Similarly, the only terms of the corresponding 3×3 traction tensor $T_{ik}^{n,s}$ that are different than zero are those that couple the vertical to the horizontal motions, i.e., the terms $T_{13}^{n,s}$, $T_{23}^{n,s}$, $T_{31}^{n,s}$, and $T_{32}^{n,s}$. This last observation suggests that if relaxed boundary conditions are assumed, i.e., the coupling of the vertical and horizontal motions is considered to be negligible, then the traction tensor can be disregarded altogether, and as a result of that Eq. (34) is reduced to

$$\tfrac{1}{2}\{u^{N,R}\} = \sum_{n=q}^{N} \sum_{s=1}^{M} [G^{n-q+1,s}]\{t^{N-n+q,s}\}, \quad R = 1, 2, \ldots, M. \qquad (40)$$

The spatial integrations indicated in Eq. (35) do not present, in general, any special difficulty and a standard numerical integration scheme, i.e., Gaussian quadrature, would produce satisfactory results. However, an improved spatial integration arrangement is used in references [5, 60] where each "source" element is divided into a number of subelements and thus the integrations indicated in Eq. (35) are performed over the area (s) of each subelement in the usual way. This further discretization within each element succeeds in taking into account in a very accurate way the influence of the wave propagation effects, even when the basic discretization scheme is rather "coarse", without increasing the computer memory requirements of the problem under consideration. The difficulty that arises in the case of singular elements, i.e., when $R = s$ in Eq. (34), can be removed with the help of a special analytical integration. Toward performing such an integration, one should first observe that in this time domain problem, due to the nature of the travelling waves, singularities can only occur during the first time step at which the tractions $\{t^{q,s}\}$ are applied over the source element s. Thus, application of Eq. (35) over the singular circular plane area defined by its center ($x_1 = 0, x_2 = 0, x_3 = 0$), and by the radius R_1 travelled by the dilatational wave during the first time interval Δt, yields in polar coordinates

$$G_{ik}^{1,s} = \frac{1}{4\pi\rho} \left\{ \frac{1}{2} \left(\frac{1}{c_2^2} - \frac{1}{c_1^2} \right) \int_0^{2\pi} \int_0^{R_2} \left(\frac{3r_i r_k}{r^2} - \delta_{ik} \right) dr\, d\theta \right.$$

$$+ \frac{1}{2} \int_0^{2\pi} \int_{R_2}^{R_1} \frac{1}{r^2} [(\Delta t)^2 - (r/c_1)^2] \left(\frac{3r_i r_k}{r^2} - \delta_{ik} \right) dr\, d\theta$$

$$\left. + \frac{1}{c_1^2} \int_0^{2\pi} \int_0^{R_1} \frac{r_i r_k}{r^2} dr\, d\theta + \frac{1}{c_2^2} \int_0^{2\pi} \int_0^{R_2} \left(\delta_{ik} - \frac{r_i r_k}{r^2} \right) dr\, d\theta \right\}, \qquad (41)$$

where R_2 is the radius travelled by the sheer wave during Δt. It is obvious that if r_i is expressed in polar coordinates as well, Eq. (41) presents no singularities.

Writing Eqs. (34) or (40) for the M elements of the discretized surface boundary, a step-by-step time marching solution of a system of $3M$ equations can be established, i.e., first for $N = q$, then for $N = q + 1$, etc., provided that $3M$ boundary conditions are available at each time step. The total number of $N(3M)^2$ discrete kernels required for this solution can be substantially reduced if the causality and time translation properties of the discretized Stokes' state are properly taken into account.

1.3.2 Frequency Domain Approach

The numerical solution of the equations related to the frequency domain BEM formulation of the elastodynamic problem appears similar but simpler than that of the time domain BEM described in the preceding section. The reason for this, of course, is the elimination of the time variable in the frequency domain formulation since the problem is transformed into a static-like one. Therefore, only a discretization in space is necessary in order to proceed with the numerical solution of the frequency domain BEM.

The basic steps of the spatial discretization procedure followed in the time domain formulation apply here as well. Thus, the two-dimensional boundary of a three-dimensional region can again be approximated by a set of rectangular elements as shown in Fig. 2. The variation of tractions and displacements over each one of those elements can, theoretically, be of any arbitrary form. However, for engineering applications, a constant variation of the related functions throughout each boundary element seems adequate [53, 54, 70, 71], and as a result, surface tractions and displacements are associated with the "nodal" point at the center of each element. On the basis of this assumption Eq. (29) can be written in a matrix summation form as

$$\tfrac{1}{2}\{\bar{u}^R\} = \sum_{s=1}^{M} \{[\bar{G}^s]\{\bar{t}^s\} - [\bar{T}^s]\{\bar{u}^s\}\}, \quad R = 1, 2, \ldots, M, \qquad (42)$$

where $\{\bar{u}^s\}$ and $\{\bar{t}^s\}$ are the Fourier transformed displacement and traction vectors, respectively, on the element s, and the $\bar{G}_{ij}^s$ and $\bar{T}_{ij}^s$ terms of the discretized fundamental solutions are, respectively, given by

$$\bar{G}_{ij}^s = \int_{(s)} \bar{u}_{ij}'(x, \xi, w)\, ds(x)$$

$$\bar{T}_{ij}^s = \int_{(s)} \bar{t}_{ij}'(x, \xi, w)\, ds(x). \qquad (43)$$

Standard numerical integration procedures can be followed in the case of non-singular elements s, i.e., $R \neq s$ in Eq. (42), while for singular elements the integrations indicated in Eq. (43) are in the sense of the Cauchy Principal Value [70] and can be estimated analytically.

With regard to the coupling of the various modes of vibration of a surface foundation, observations similar to those made earlier for the time domain approach are also true in this case. Thus, if relaxed boundary conditions are considered, the influence of the fundamental traction tensor vanishes and Eq. (42) is simplified to

$$\tfrac{1}{2}\{\bar{u}^R\} = \sum_{s=1}^{M} [\bar{G}^s]\{\bar{t}^s\}, \quad R = 1, 2, \ldots, M. \tag{44}$$

In both the time and frequency domain SSI techniques considered in this work [5, 60, 53, 54], the fundamental solutions corresponding to the three-dimensional infinite space are used. In doing so, the infinite free surface of the half-space, being the natural boundary of the surface foundation problem, needs to be discretized when complete bond is assumed between the foundation and the half-space. However, parametric studies conducted by Karabalis and Mohammadi [79] and Dominguez [53, 54] indicate that only few elements on the free surface surrounding the contact surface between the half-space and the foundation are necessary for the results to be satisfactory. In the case of relaxed boundary conditions, of course, only the contact surface needs to be discretized since the boundary condition on the free surface of the half-space requires zero tractions to be applied. Also, should half-space fundamental solutions in time or frequency domain [72, 73] be used, there would be no need for discretizing the free surface under either relaxed or non-relaxed boundary conditions. Half-space fundamental solutions, however, are not so popular because of their complexity.

1.4 Massless Foundations

In this section, the Boundary Element Methodologies reviewed in the preceding part of the article will be applied to SSI problems involving massless, rigid, surface foundations resting on a linear elastic, homogeneous, and isotropic half-space. It will be shown later that once this kinematic problem has been solved the mass of the foundation and of the superstructure can be taken into account in a straight-forward manner on the basis of the results obtained in this section.

In order to proceed with the solution of SSI problems in the mainframe of a substructure methodology [33, 39], two more sets of equations are required in addition to the BIE for the time domain, i.e., Eq. (34) or (40), or for the frequency domain, i.e., Eq. (42) or (44); namely: (i) the compatibility equations, and (ii) the equilibrium equations, written in terms of the displacements and the tractions, respectively, on the contact surface between the soil and the foundation.

The compatibility condition for the displacements on the contact surface under a rigid surface foundation during time step N can be expressed as

$$\{u^N\} = [S]\{D^N\}, \tag{45}$$

where $\{u^N\}$ is the $(3M \times 1)$ displacement vector corresponding to the centers of the

M boundary elements and $\{D^N\}$ is the (6×1) rigid body displacement vector given by

$$\{D^N\} = [\Delta_1^N \Delta_2^N \Delta_3^N \phi_1^N \phi_2^N \phi_3^N]^T, \tag{46}$$

with Δ_i^N and ϕ_i^N being the rigid body displacements and small rotations, respectively, with respect to a system of coordinate axes attached to the center of the foundation as shown in Fig. 2a. The $(3M \times 6)$ transformation matrix $[S]$ is the assemblage of the M submatrices $[S_m]$ expressing the displacement vector at the center of each element m $(m = 1, 2, \ldots M)$ due to a unit rigid body motion, and can be written as

$$[S_m] = \begin{bmatrix} 1 & 0 & 0 & 0 & 0 & -x_2^m \\ 0 & 1 & 0 & 0 & 0 & x_1^m \\ 0 & 0 & 1 & x_2^m & -x_1^m & 0 \end{bmatrix}, \tag{47}$$

where x_i^m are the coordinates of the center of the element m, with $x_3^m = 0$.

A Fourier transform expression equivalent to Eq. (45) can be obtained by substituting the time dependent displacement functions with frequency dependent displacement amplitudes, thus

$$\{\bar{u}\} = [S]\{\bar{D}\}, \tag{48}$$

where the transformation matrix $[S]$ is the same as in the time domain case.

With regard to the dynamic disturbances acting on the foundation, two distinct cases should be examined: (i) externally applied loads, and (ii) obliquely incident seismic waves. For purposes of clarity these two cases are examined separately in this work, although unified approaches have also been presented elsewhere, e.g., [8].

1.4.1 Externally Applied Dynamic Loads

The equilibrium of external forces and contact tractions applied on a massless foundation during time step N can be written in a matrix form as

$$\{P^N\} = [K]\{t^N\}, \tag{49}$$

where

$$\{P^N\} = [P_1^N P_2^N P_3^N M_1^N M_2^N M_3^N]^T, \tag{50}$$

are the external forces P_i and moments M_i along and about the axes x_i of Fig. 2a, and the traction vector $\{t^N\}$ consists of an M number of (3×1) vectors $\{t_m^N\}$ representing the traction vector at the center of the element m $(m = 1, 2, \ldots, M)$. The $(6 \times 3M)$ transformation matrix $[K]$ is assembled by an M number of (6×3) submatrices $[K_m]$ $(m = 1, 2, \ldots, M)$ given by

$$[K_m] = A_m[S_m]^T, \tag{51}$$

with A_m being the area of the element m.

Equations (34), (45) and (49) can be viewed together as a system of linear algebraic equations with unknowns being the rigid body displacements and the contact tractions during time step N. The solution of this system of equations can be accomplished by a step-by-step time marching, i.e., for $N = q$, then for $N = q + 1$, etc. If relaxed boundary conditions are used Eq. (34) is replaced by Eq. (40). The solution to this last case is given in Refs. [5, 60].

A frequency domain equivalent of Eq. (49) can be obtained by substituting the time dependent external forces and tractions by their frequency counterparts

$$\{\bar{P}\} = [K]\{\bar{t}\},\tag{52}$$

where the matrix $[K]$ remains the same as before. Then, as in the time domain case, Eqs. (42), (48) and (52) form a system of algebraic equations with the unknowns being the frequency dependent rigid body displacements and contact tractions. The solution to this system of equations will result in a force-displacement relationship of the form

$$\{\bar{P}\} = [K(w)]\{\bar{D}\}\tag{53}$$

where $[K(w)]$ is the (6×6) impedance matrix of a rigid surface foundation.

1.4.2 Obliquely Incident Seismic Waves

When an elastic half-space is excited by a plane incident wave, as shown in Fig. 3, the displacement vector field $\{u_g\}$ on the free surface $x_3 = 0$ is described, in the absence of any foundation, by the general expression

$$\{u_g\} = \{u_g(x, t)\} = [u_{gx_1} u_{gx_2} u_{gx_3}]^T,\tag{54}$$

where $\{u_g(x, t)\}$ is dependent upon the vertical and horizontal angles of incidence, θ_v and θ_H, respectively, as well as the apparent velocity $c = c_1/\cos\theta_v$ or $c = c_2/\cos\theta_v$, depending on the type of the incident wave. In the case that a foundation is placed on the surface of the half-space, a scattered displacement field $\{u_s(x, t)\}$ is generated in addition to the free field displacements. Following Thau's work [74] the total displacement field $\{u(x, t)\}$ occurring under the foundation will be given as the summation of the free and the scattered displacement fields

$$\{u(x, t)\} = \{u_g(x, t)\} + \{u_s(x, t)\}.\tag{55}$$

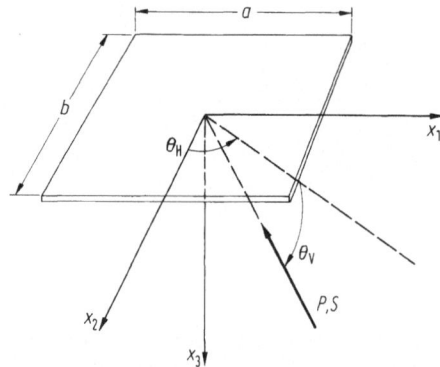

Fig. 3. Surface foundation subjected to obliquely incident P and S waves (after Karabalis [5])

Both the time and the frequency domain formulations of the surface foundation problem reviewed in this work proceed on the basis of Eq. (55).

Along the line of the time domain procedure developed in the previous sections of this article, the displacement functions involved in Eq. (55) should be discretized in time and space. For the time discretization the rectangular impulse approximation shown in Fig. 1 can again be applied. Thus, the functions $u_g(x, t)$ and $u_s(x, t)$ can be represented by a sequence of rectangular impulses as

$$\{u_g(x, t)\} = \{u_g^q(x)\} \{H[t - (q - 1)\Delta t] - H(t - q\Delta t)\}$$
$$\{u_s(x, t)\} = \{u_s^q(x)\} \{H[t - (q - 1)\Delta t] - H(t - q\Delta t)\},$$

(56)

where $\{u_g^q(x)\}$ and $\{u_s^q(x)\}$ are the intensities of the rectangular impulses $u_g(x, t)$ and $u_s(x, t)$, respectively, during the time step q. As far as the spatial discretization is concerned the contact surface is approximated by M rectangular elements, and a constant variation of displacements is assumed over each one of them. Under these assumptions Eq. (55) can finally be written for the total number of M elements and the time step q as

$$\{u^q\} = \{u_g^q\} + \{u_s^q\},$$

(57)

where each of the matrices $\{u^q\}$, $\{u_g^q\}$, and $\{u_s^q\}$ contain an M number of (3×1) vectors u_i^q, u_{gi}^q and u_{si}^q, expressing the total, the free field, and the scattered field displacement vectors, respectively, at the center of the element i. With regard to the various quantities involved in Eq. (57) one can observe that: (1) the total displacement field $\{u^q\}$ can be expressed in terms of the rigid body displacements, for the case of a rigid foundation, by using the restraint Eq. (45), (ii) the scattered displacement field $\{u_s^q\}$ is due to the rigid body motion of the foundation and is given in terms of the contact tractions by Eq. (34) or Eq. (40), and iii) the free field displacement vector $\{u_g^q\}$ which could have any time variation is assumed to be known for the particular problem at hand. Thus, Eq. (34) or (40), depending on whether or not relaxed boundary conditions are used, along with Eqs. (45), (49), and (57), form a system of linear algebraic equations with the unknowns being the rigid body displacements and contact tractions. A step-by-step time marching solution of this system of equations is given in Refs. [5, 60] for relaxed boundary conditions.

In the Fourier transformed domain formulation, instead of the general expressions (54) and (55) their frequency dependent harmonic counterparts can be used. Then on the basis of a spatial discretization identical to the one applied in the time domain formulation, the transformed domain equivalent of Eq. (57) is obtained as

$$\{\bar{u}\} = \{\bar{u}_g\} + \{\bar{u}_s\},$$

(58)

where each of the matrices $\{\bar{u}\}$, $\{\bar{u}_g\}$, and $\{\bar{u}_s\}$ contains an M number of (3×1) vectors $\bar{u}_i$, $\bar{u}_{gi}$, and $\bar{u}_{si}$, expressing the total, the free field, and the scattered field harmonic displacement vectors, respectively, at the center of the element i ($i = 1, 2, \ldots, M$). For some fixed value of the frequency w, Eqs. (42), or (44) if relaxed boundary conditions are used, (48), (52), and (58) can be viewed together as a system of linear algebraic equations with the unknowns being the rigid body displacements and contact tractions of the Fourier transformed domain. The solution of this system of equations, in absence of external forces, will result in a relationship of the form

$$\{\bar{D}\} = [C(w)]\{\bar{u}_g\}, \tag{59}$$

where $[C(w)]$ is the $(6 \times 3M)$ input motion matrix of the rigid surface foundation. An extensive discussion on the importance and physical meaning of the input motion matrix can be found in Ref. [27].

1.5 Massive Foundations – Superstructure

Once the response of a massless rigid foundation is computed, as it was done, for example, in the previous section, the dynamics of a massive superstructure supported on that foundation, which is now assumed to possess a mass, can be studied with the help of the superposition principle for linear soil-structure interaction analysis [6]. According to this approach the motion of a massive structure subjected to seismic wave excitation can be analyzed in two steps, Fig. 4. In the first step, also called "kinematic interaction", the response of a weightless structure, which is otherwise identical to the original structure, is computed due to the dynamic disturbance under consideration. In the second step, also called "inertial interaction", the accelerations computed in the first step are multiplied by the corresponding masses, and the response of the structure-soil system to these fictitious inertia forces is computed. The total response of the initial structure can then be found by adding the responses obtained in these two steps. The mathematical validity of this superposition principle can be proven by a simple partition of the equations of motion and is given in various sources, e.g. [6, 75, 76]. It should be noted at this point that if the structure is laying on a rigid foundation the stiffness of the structure need not be considered in the first step of the analysis. The application of the SSI superposition principle in connection with the time domain formulation described in the previous sections of this article has not been reported as yet.

A similar approach in a transformed domain formulation is presented in Ref. [31] and is used here for illustrative purposes. The rigid body displacements $\{\bar{D}_K\}$ of a massless surface foundation due to a free field wave excitation $\{\bar{u}_g\}$ can be obtained through an application of Eq. (59) as

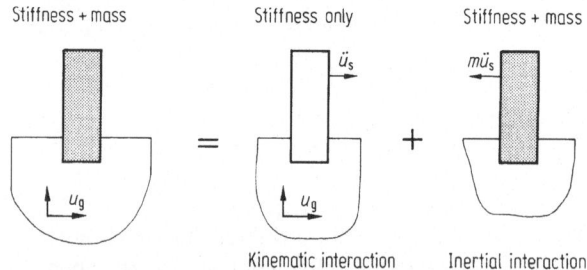

Stiffness + mass   Stiffness only   Stiffness + mass

Kinematic interaction   Inertial interaction

Fig. 4. Soil-structure interaction superposition principle

$$\{\bar{D}_K\} = [C(w)]\{\bar{u}_g\}. \tag{60}$$

This is the first step of the superposition principle. The fictitious inertia forces $\{\bar{F}_I\}$ corresponding to the total rigid foundation displacements $\{\bar{D}_T\}$ can be written as

$$\{\bar{F}_I\} = w^2[[M_f] + [M_s(w)]]\{\bar{D}_T\}, \tag{61}$$

where $[M_f]$ is the (6×6) mass matrix of the foundation, and $[M_s(w)]$ is a (6×6) frequency dependent equivalent mass matrix which accounts for the elastic properties and the mass distribution of the superstructure. The response $\{\bar{D}_I\}$ of the rigid foundation to these inertia forces, i.e., the "inertial interaction", can be found by applying Eq. (53)

$$\{\bar{D}_I\} = [K(w)]^{-1}\{\bar{F}_I\}. \tag{62}$$

Finally, the total response $\{\bar{D}_T\}$ at the foundation level can be found by adding Eqs. (60) and (62) by parts, i.e.,

$$\{\bar{D}_T\} = \{\bar{D}_K\} + \{\bar{D}_I\} = [[I] - w^2[K(w)]^{-1}[[M_f] + [M_s(w)]]]^{-1}[C(w)]\{\bar{u}_g\}, \tag{63}$$

where $[I]$ is the identity matrix. The response at any level of the structure can be obtained by standard techniques, i.e., through a FEM discretization of the structure, in terms of the total motion at the foundation level. It should be noted at this point that the approach followed in Refs. [5, 62] for coupling the BEM with the FEM at the soil-structure interface can also be used to study the response of the complete soil-foundation-structure system thus avoiding the intermediate step that requires the use of the equivalent mass matrix $[M_s(w)]$ in Eq. (61).

1.6 Numerical Results

The BEM methodologies described in the previous sections have been applied to a variety of problems in the general area of SSI. In this section a comparison study is attempted among the time domain BEM of Refs. [5, 60], the frequency domain BEM of Refs. [53, 54], and the frequency domain methodology of Ref. [19]. Since no results have been presented by either of the above mentioned frequency domain methodologies in the time domain, this comparison study deals with harmonic disturbances only. Also given, however, are some results which demonstrate the capability of the time domain BEM to handle any time variation of the dynamic forcing function. For the sake of brevity the examples presented in this section are limited to square, rigid, surface, massless foundations subjected to externally applied dynamic forces.

As a first example, the vertical, horizontal, and rocking impulse responses of a 5×5 ft footing, as computed by the time domain BEM through a direct step-by-step integration, are given in Fig. 5a–c. These responses are due to an externally applied rectangular impulse vertical force or horizontal force or overturning moment described, respectively, by

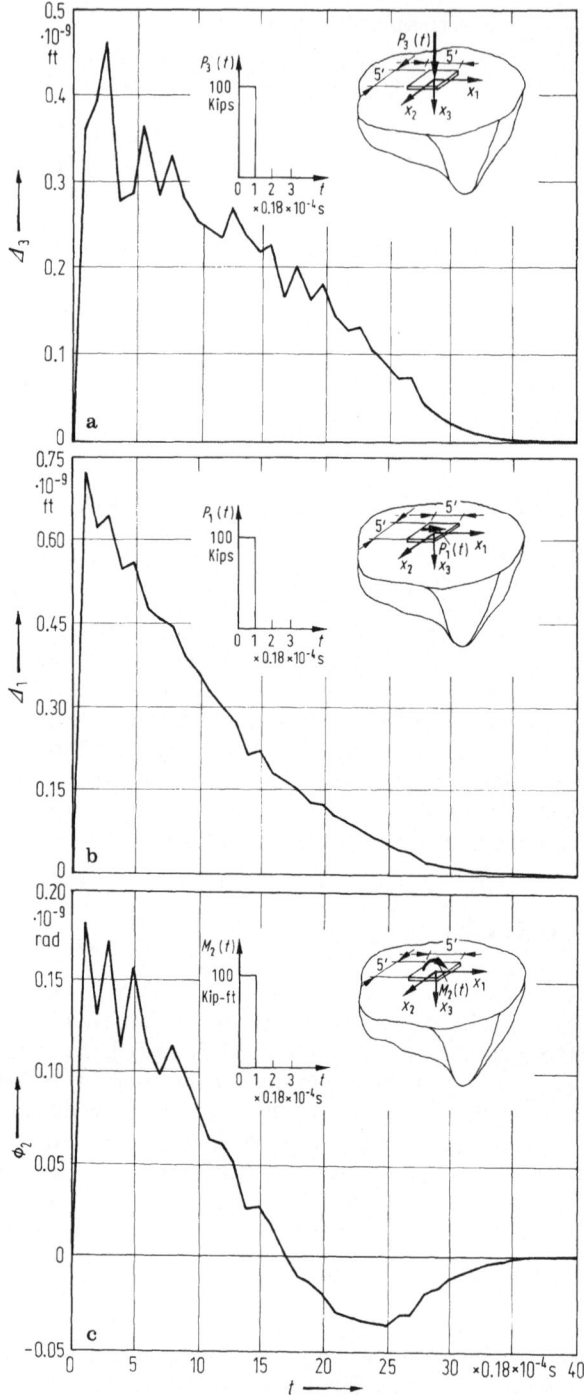

Fig. 5a–c. Vertical, horizontal and rocking impulse responses of a square foundation (after Karabalis [5])

$$P_3 = P_1 = M_2 = \begin{cases} 180, & \text{first time step} \\ 0, & \text{elsewhere,} \end{cases} \tag{64}$$

where the units for the magnitudes P_3 and P_1 are kips and for the M_2 are kip-ft. Relaxed boundary conditions are assumed for all three motions, i.e., the shear contact stresses $\sigma_{13} = \sigma_{23} = 0$ for the vertical and rocking motions, and the vertical and horizontal contact stresses $\sigma_{33} = \sigma_{23} = 0$ for the horizontal motion along the x_1-axis. The results shown in Fig. 5a–c have been obtained by using an 8×8 element basic discretization scheme with a secondary 5×5 subelement mesh per element. The time step Δt is related to the size of the elements used and is given by

$$\Delta t = (1/c_1)\sqrt{(\Delta a \cdot \Delta b)/\pi}, \tag{65}$$

where Δa and Δb are the lengths of the sides of a typical rectangular element within the discretization scheme under consideration. The half-space of this example is characterized by a modulus of elasticity $E = 2.58984 \times 10^9$ lb/ft^2, a Poisson's ratio $\nu = \frac{1}{3}$ and a mass density $\rho = 10.368$ lb$\cdot$s^2/ft^4. It should be noted at this point, however, that once the impulse responses given in Fig. 5a–c have been obtained, they can be used in a convolution approach to compute the response of the given foundation to any generally time-varying forcing function. More details about this superposition approach, which is applicable to linear systems only, can be found in references [5, 67].

As a second example, the vertical compliance function $C_{VV}(a_0)$, the horizontal compliance function $C_{HH}(a_0)$, and the rocking compliance function $C_{MM}(a_0)$ of a square, rigid, surface, massless foundation are plotted, in Fig. 6a–c, versus the dimensionless frequency a_0. For the purposes of this article the compliance functions are defined as

$$C_{VV}(a_0) = \mu b \Delta_3 / P_3$$
$$C_{HH}(a_0) = \mu b \Delta_1 / P_1 \tag{66}$$
$$C_{MM}(a_0) = \mu b^3 \phi_2 / M_2,$$

where $2b$ is the length of the side of the square foundation, Δ_3, Δ_1, and ϕ_2 are the amplitudes of the rigid body vertical displacement, horizontal displacement along the x_1-axis, and rotation about the x_2-axis, respectively, and similarly, P_3, P_1, and M_2 are the amplitudes of the applied vertical force, horizontal force along the x_1-axis, and overturning moment about the x_2-axis, respectively. The dimensionless frequency is given by

$$a_0 = wb/c_2. \tag{67}$$

The results shown in Fig. 6a–c have been obtained using relaxed boundary conditions and an 8×8 element discretization mesh by each method. The close agreement of the results produced by all three methods is apparent. Furthermore, it is shown in Refs. [53, 79] that results generated by using a complete bond between the foundation and the half-space are similar to those obtained under the assumption of relaxed boundary conditions.

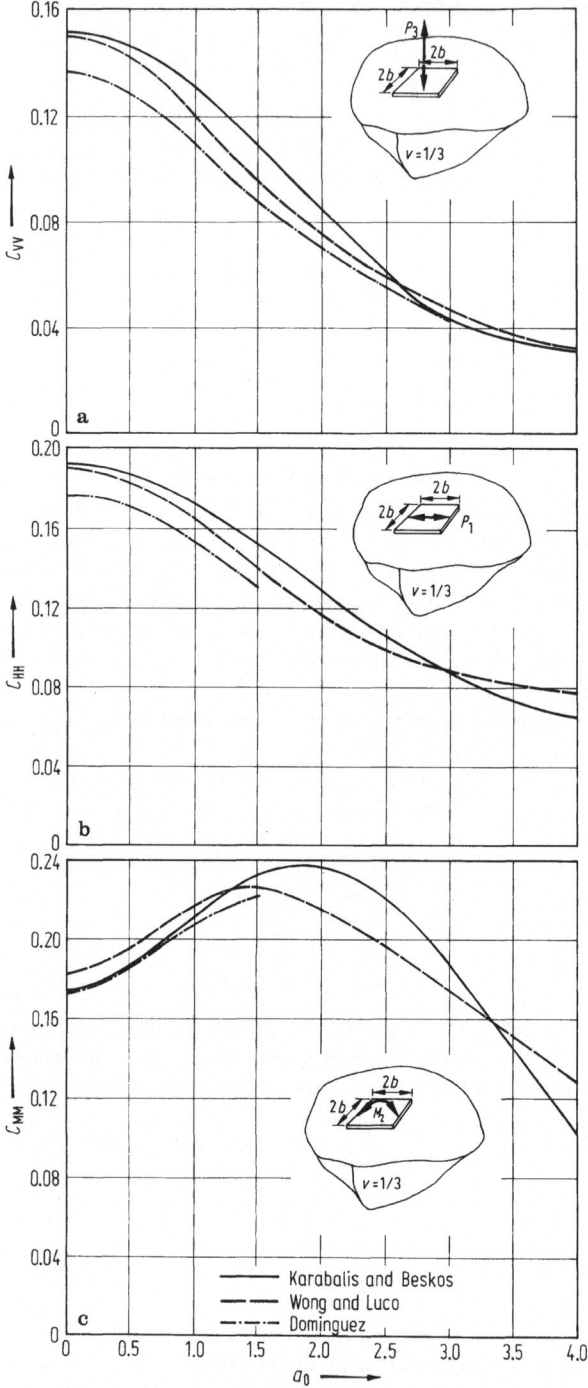

Fig. 6a–c. Vertical, horizontal and rocking compliances of a square foundation

1.7 Conclusions

The application of two general BEM methodologies, namely, the time domain approach and the Fourier transform domain approach, to some representative dynamic SSI interaction problems has been examined in this article. A comparison of these surface gridding methods with other well established volume gridding methods, i.e., the FEM or the FDM, shows that for SSI problems, which by nature involve infinite domains, the BEM has the advantages of: (i) requiring a minimum amount of discretization and thus being very efficient, and (ii) taking into account automatically the radiation condition which results in an improved accuracy. However, since no applications of the BEM to problems involving general soil inhomogeneities and nonlinearities have yet been reported, one has to resort to either large and expensive FEM or FDM models, or to hybrid methods, i.e., a BEM-FEM combination, for such analyses.

In comparing the time domain BEM presented in this article to frequency domain methodologies in general, one can observe that: (i) the time domain approach is a direct and natural way of studying wave propagation phenomena in a time step-by-step fashion, in contrast to the static-like frequency domain solutions where the time variable has been eliminated, (ii) use of the time domain formulation provides a solution to elastodynamic problems in one direct step, whereas frequency domain formulations ask for a two step solution, i.e., transformation and solution of the problem in the frequency domain and then reverse transformation into time domain, and (iii) the time domain approach forms the basis for an extension to SSI problems involving nonlinearities in the structure and/or the soil medium, which is not possible by the frequency domain methods. Both the time and the frequency domain Boundary Element Methods reviewed in this article have been reported to possess a high degree of accuracy and efficiency, and to have excellent convergence characteristics, even when coarse discretization schemes are utilized.

References

1 Reissner, E., "Stationäre, axialsymmetrische, durch eine schüttelnde Masse erregte Schwingungen eines homogenen elastischen Halbraumes", Ing. Arch., 7, 381–396 (1936).

2 Reissner, E., "Freie und erzwungene Torsionschwingungen des elastischen Halbraumes", Ing. Arch., 8, 229–245 (1937).

3 Richart, F.E., Jr., Hall, J.R., Jr. and Woods, R.D., "Vibrations of soils and foundations", Prentice Hall, Englewood Cliffs, N.J., 1970.

4 Gazetas, G., "Analysis of machine foundation vibrations: state of the art", Soil Dyn. Earth. Engng., 2, 2–42 (1983).

5 Karabalis, D.L., "Dynamic response of three-dimensional foundations", Ph.D. Thesis, University of Minnesota, Minnespolis, Minnesota, 1984.

6 Seed, H.B., Whitman, R.V. and Lysmer, J., "Soil-structure interaction effects in the design of nuclear power plants", pp. 220–241 in Hall, W.J., Ed., "Structural and Geotechnical Mechanics", Prentice Hall, Englewood Cliffs, N.J., 1977.

7 Novak, M., "Foundation and soil-structure interaction", pp. 1421–1448, in Proc. 6th World Conf. Earth. Engng., N. Delhi, India, 1977.

8 Luco, J.E., "Linear soil-structure interaction: a review", pp. 41–57 in Datta, Ed., "Transient motions in earthquake engineering", ASME, 1982.

9 Collins, W.D., "The forced torsional oscillations of an elastic half-space and an elastic stratum", Proc. London Math. Soc., *12*, 226–244 (1962).

10 Robertson, I.A., "Forced vertical vibration of a rigid circular disc on a semi-infinite elastic solid", Proc. Campr. Phil. Soc., *62*, 547–553 (1966).

11 Gladwell, G.M.L., "Forced tangential and rotatory vibration of a rigid circular disc on a semi-infinite solid", Int. J. Engng. Sci., *6*, 591–607 (1968).

12 Veletsos, A.S. and Verbic, B., "Vibration of viscoelastic foundations", Earth. Engng. Struct. Dyn., *2*, 87–102 (1973).

13 Veletsos, A.S. and Wei, Y.T., "Lateral and rocking vibration of footings", Proc. ASCE, *97*, SM9, 1227–1248 (1971).

14 Luco, J.E. and Westmann, R.A., "Dynamic response of circular footings", Proc. ASCE, *97*, EM5, 1381–1395 (1971).

15 Luco, J.E. and Westmann, R.A., "Dynamic response of a rigid footing bonded to an elastic half-space", J. Appl. Mech., *39*, 527–534 (1972).

16 Lamb, H., "On the propagation of tremors over the surface of an elastic solid", Phil. Trans. Roy. Soc., London, *A203*, 1–42 (1904)

17 Lysmer, J., "Vertical motion of rigid footings", Ph.D. dissertation, Univ. of Michigan, Ann Arbor, Michigan, August 1965.

18 Elorduy, J., Nieto, J.A. and Szekely, E.M., "Dynamic response of bases of arbitrary shape subjected to periodic vertical loading", pp. 105–121, Proc. Int. Symp. Wave Propagation and Dynamic Properties of Earth Materials, Albuquerque, New Mexico, August 1967.

19 Wong, H.L. and Luco J.E., "Dynamic response of rigid foundations of arbitrary shape", Earth. Engng. Struct. Dyn., *4*, 579–587 (1976).

20 Gaul, L., "Dynamische Wechselwirkung eines Fundamentes mit dem viscoelastischen Halbraum", Ing. Arch., *46*, 401–422 (1977).

21 Kitamura, Y. and Sakurai, S., "Dynamic stiffness for rectangular rigid foundations on a semi-infinite elastic medium", Int., J. Num. Anal. Meth. Geomech., *3*, 159–171 (1979).

22 Kitamura, Y. and Sakurai, S., "A numerical method for determining dynamic stiffness", pp. 393–399, Eisenstein, Z., Ed., "Numerical Methods in Geomechanics, Edmonton 1982", Balkema, Rotterdam, 1982.

23 Adeli, H., Hejazi, M.S., Keer, L.M. and Nemat-Nasser, S., "Dynamic response of foundations with arbitrary geometries", Proc. ASCE, *107*, EM5, 953–967 (1981).

24 Hamidzadeh-Eraghi, H.R. and Grootenhuis, P., "The dynamics of a rigid foundation on the surface of an elastic half-space", Earth. Engng. Struct. Dyn., *9*, 501–515 (1981).

25 Savidis, S.A. and Richter, T., "Dynamic interaction of rigid foundations", pp. 369–374, Proc. 9th Int. Conf. Soil Mech. Found. Engng., Vol. 2, Tokyo, Japan, 1977.

26 Gantayat, A. and Kamil, H., "An impedance function approach for soil-structure interaction analyses including structure-to-structure interaction effects", Trans. 6th SMiRT, Paris, August 1981.

27 Wong, H.L. and Luco, J.E., "Dynamic response of rectangular foundations in obliquely incident seismic waves", Earth. Engng. Struct. Dyn., *6*, 3–16 (1978).

28 Luco, J.E. and Wong, H.L., "Dynamic response of rectangular foundations for Rayleigh wave excitation", pp. 1542–1548, Proc. 6th World Conf. Earth. Engng, N. Delhi, India, 1977.

29 Bielak, J. and Coronato, J.A., "Response of multiple-mass systems to nonvertically incident seismic waves", pp. 801–804, Proc. Int. Cont. Recent Advances in Geotech. Earth. Engng. and Soil Dyn., St. Louis, Univ. of Missouri-Rolla, 1981.

30 Werner, S.D., Lee, L.C., Wong, H.L. and Trifunac, M.D., "Structural response to travelling seismic waves", Proc. ASCE, *105*, ST 12, 2547–2564 (1979).

31 Luco, J.E. and Wong, H.L., "Response of structures to nonvertically incident seismic waves", Bull. Seism. Soc. Amer., *72*, 275–302 (1982).

32 Roesset, J.M. and Gonzalez, J.J., "Dynamic interaction between adjacent structures", pp. 127–166, Vol. 1, Prange, B., Ed., "Dynamical Methods in Soil and Rock Mechanics", Balkema, Rotterdam, 1978.

33 Gupta, S., Penzien, J., Lin, T.W. and Yeh, C.S., "Three-dimensional hybrid modeling of soil-structure interaction", Earth. Engng. Struct. Dyn. *10*, 69–87 (1982).

34 Dasgupta, S.P. and Rao, N.S.V.K., "Dynamics of rectangular footings by finite elements", Proc. ASCE, *104*, GT5, 621–637 (1978).

35 Roesset, J.M. and Ettouney, M.M., "Transmitting boundaries: a comparison", Int. J. Num. Anal. Meth. Geomech., *1*, 151–176 (1977).

36 Kausel, E. and Tassoulas, J.L., "Transmitting boundaries: a closed form comparison", Bull. Seism. Soc. Amer., *71*, 143–159 (1981).

37 Bettess, P. and Zienkiewicz, O.C., "Diffraction and refraction of surface waves using finite and infinite elements", Int. J. Num. Meth. Engng., *11*, 1271–1290 (1977).

38 Chow, Y.K. and Smith, I.M., "Infinite elements for dynamic foundation analysis", pp. 15–22, Vol. 1, Eisenstein, Z., Ed., "Numerical Methods in Geo-Mechanics, Edmonton 1982", Balkema, Rotterdam, 1982.

39 Gutierrez, J.A. and Chopra, A.K., "A substructure method for earthquake analysis of structures including structure-soil interaction", Earth. Engng. Struct. Dyn., *6*, 51–69 (1978).

40 Chopra, A.K., Chakrabarti, P. and Dasgupta, G., "Dynamic stiffness matrices for viscoeleastic half-plane foundations" Proc. ASCE, *102*, EM3, 497–514 (1976) and Proc. ASCE, *105*, EM5, 729–745 (1979).

41 Nelson, I. and Isenberg, J., "Soil island approach to structure-media interaction", pp. 41–57, in Desai, C.S., Ed., "Numerical Methods in Geomechanics", ASCE, N.Y., 1976.

42 Day, S.M. and Frazier, G.A., "Seismic response of hemisperical foundation", Proc. ASCE, *105*, EM1, 29–41 (1979).

43 Dasgupta, G., "Foundation impedance matrices for embedded structures by substructure deletion", Proc. ASCE, *106* EM3, 517–523 (1980).

44 Dasgupta, G., "A finite element formulation for unbounded homogeneous continua", J. Appl. Mech., *49*, 136–140 (1982).

45 Murakami, H., Shioya, S., Yamada, R. and Luco, J.E., "Transmitting boundaries for time-harmonic elastodynamics on infinite domains", Int. J. Num. Meth. Engng., *17*, 1697–1716 (1981).

46 Hadjian, A.H., Luco, J.E. and Tsai, N.C., "Soil-structure interaction: continuum or finite element?", Nucl. Engng. Des., *31*, 151–167 (1974).

47 Seed, H.B., Lysmer, J. and Hwang, R., "Soil-structure interaction analyses for seismic response", Proc. ASCE, *101*, GT5, 439–457 (1975).

48 Ang, A.H.S. and Newmark, N.M., "Computation of underground structural response", Univ. of Illinois Report for Defense Atomic Support Agency (now Defense Nuclear Agency), DASA Rept. 1386, Washington, D.C., June 1963.

49 Wilkins, M.L., et al., "A method for computer simulation of problems in solid mechanics and gas dynamics in three dimensions and time", Report UCRL-51574, Lawrence Livermore Laboratory, Univ. of California, 1974.

50 Robinson, A.R., "The transmitting boundary-again", pp. 163–177, in Hall, W.J., Ed., "Structural and Geotechnical Mechanics", Prentice Hall, Englewood Cliffs, N.J., 1977.

51 Cundal, P.A., Kunar, R.R., Carpenter, P.C. and Marti, J., "Solution of infinite dynamic problems by finite modeling in the time domain", Proc. 2nd Int. Conf. on Applied Numerical Modelling, Madrid, Spain, Sept. 11–18, 1978.

52 Kunar, R.R. and Rodriguez-Ovefero, L., "A model with non-reflecting boundaries for use in explicit soil-structure interaction analyses", Earth. Engng. Struct. Dyn., *8*, 361–374 (1980).

53 Dominguez, J., "Dynamic stiffness of rectangular foundations", Publ. No. R78-20, Dept. of Civil Engng., M.I.T., August 1978.

54 Dominguez, J., "Response of embedded foundations to travelling waves", Publ. No. R78-24, Dept. of Civil Engng., M.I.T., August 1978.

55 Ottenstreuer, M. and Schmid, G., "Boundary elements applied to soil-foundation interaction", pp. 293–309 in Proc. 3rd Int. Sem. on Recent Advances in Boundary Element Methods, Irvine, Calif., July 1981.

56 Ottenstreuer, M., "Frequency dependent dynamic response of footings", pp. 799–809 in Proc. Soil Dynamics and Earth. Engng. Conf., Southampton, England, July 1982.

57 Apsel, R.J., "Dynamic Green's functions for layered media and applications to boundary value problems", Ph.D. Thesis, Univ. of California, San Diego, Calif., 1979.

58 Wolf, J.P. and Darbre, G.R., "Dynamic-stiffness matrix of soil by the boundary-element method: conceptual aspects", Earth. Engng. Struct. Dyn., *12*, 385–400 (1984).

59 Wolf, J.P. and Darbre, G.R., "Dynamic-stiffness matrix of soil by the boundary-element method: embedded foundation", Earth. Engng. Struct. Dyn., *12*, 401–416 (1984).

60 Karabalis, D.L. and Beskos, D.E., "Dynamic response of 3-D rigid surface foundations by time domain boundary element method", Earth. Engng. Struct. Dyn., *12*, 73–93 (1984).

61 Karabalis, D.L. and Beskos, D.E., "Earthquake response of 3-D foundations by the boundary element method", pp. 769–776, Proc. 8th World Conf. Earth. Engng., San Francisco, U.S.A., 1984.

62 Karabalis, D.L. and Beskos, D.E., "Dynamic response of 3-D flexible foundations by time domain BEM and FEM", Int. J. Soil Dyn. Earth. Engng., Vol. 4, pp. 91–101, 1985.

63 Zienkiewicz, O.C., Kelly, D.W. and Bettess, P., "The coupling of finite element and boundary solution procedures", Int. J. Num. Meth. Engng., *11*, 355–376 (1977).

64 Kelly, D.W., Mustoe, G.G.W. and Zienkiewicz, O.C., "Coupling boundary element methods with other numerical methods", pp. 251–285, Banerjee, P.K. and Butterfield, R., Eds., Development in Boundary Element Methods-1, Applied Science Publishers, Ltd., 1979.

65 Varadarajan, A. and Singh, R.B., "Analysis of tunnels by coupling FEM with BEM", pp. 611–618, Eisenstein, Z., Ed., Numerical Methods in Geomechanics Edmonton 1982, Balkema, Rotterdam, 1982.

66 Mita, A. and Takanashi, W., "Dynamic soil-structure interaction analysis by hybrid method", pp. 785–794, Brebbia, C.A., Futagami, T. and Tanaka, M., Eds., Proc. 5th Int. Conf. in Boundary Elements, Hiroshima, Japan, 1983.

67 Veletsos, A.S. and Verbic, B., "Basic response functions for elastic foundations", J. Eng. Mech. Div., ASCE, *100*, 189–202 (1974).

68 Eringen, A.C. and Suhubi, E.S., "Elastodynamics-Vol. II, Linear Theory", Academic Press, New York, 1975.

69 Cruse, T.A. and Rizzo, F.J., "A direct formulation and numerical solution of the general transient elastodynamic problem I", J. Math. Anal. Appl., *22*, 244–259 (1968).

70 Cruse, T.A., "A direct formulation and numerical solution of the general transient elastodynamic problem II", J. Math. Anal. Appl., *22*, 341–355 (1968).

71 Manolis, G.D. and Beskos, D.E., "Dynamic stress concentration studies by boundary integrals and Laplace transform", Int. J. Numer. Methods Eng., *17*, 573–599 (1981).

72 Johnson, L.R., "Green's function for Lamb's problem," Geophys. J.R. Astr. Soc., *37*, 99–131 (1974).

73 Kobayashi, S. and Nishimura, N., "Green's tensors for elastic half-spaces. – An application of Boundary Integral Equation Method", Memoirs of the Faculty of Engineering, Kyoto University, *42*, part 2, 228–241 (1980).

74 Thau, S.A., "Radiation and scattering from a rigid inclusion in an elastic medium", J. Appl. Mech., ASME, *34*, 509–511 (1967).

75 Kausel, E., Whitman, R.V., Morray, J.P. and Elsabee, F., "The spring method for embedded foundations", Nucl. Engng. Des., *48*, pp. 377–392 (1978).

76 Whitman, R.V. and Bielak, J., "Foundations", pp. 223–260, Rosenblueth, E., Ed., Design of Earthquake Resistant Structures, John Wiley & Sons, New York, 1980.

77 Karabalis, D.L. and Beskos, D.E., "Dynamic Analysis of 3-D rigid embedded foundations by time domain boundary element method," Computer Methods in Applied Mechanics and Engineering Vol. 56, pp. 91–119, 1986.

78 Gaitanaros, A.P. and Karabalis, D.L., "3-D flexible embedded machine foundations by the BEM and the FEM," pp. 81–96, in: Karabalis, D.L., Ed., "Recent Applications in Computational Mechanics," ASCE, New York, 1986.

79 Karabalis, D.L. and Mohammadi, M., "The application of the Boundary Element Method to dynamic soil-structure interaction problems: Computational aspects," pp. 321–328, Proceedings of the XIII Southeastern Conference on Theoretical and Applied Mechanics, Columbia, South Carolina, April 1986.

80 Karabalis, D.L. and Beskos, D.E., "Dynamic soil-structure interaction," Chapter 11, in Beskos D.E., Ed., "Boundary Element Methods in Mechanics," North-Holland, Amsterdam, 1987.

Chapter 2

Dynamics of Foundations

by J. Dominguez and R. Abascal

2.1 Introduction

Dynamics of foundations is part of the more general field dynamic soil-structure interaction, which is concerned with the study of structures based on flexible soils and subjected to dynamic actions that may be directly applied to the structure or transmited through the soil.

In its early stages, the dynamic soil-structure interaction studied problems with external excitation; in most cases vibrations of machines on elastic foundations. Tall buildings, or any other structure based on the ground, subjected to wind loads, are also examples of problems with external excitation. More recently, important structures excited by the motion of the soil are being studied; for instance, masive or tall structures in seimic areas, and underground structures resistant to earth-quake and blast loads. In all of the above examples, the flexibility of the soil has an important effect on the response of the structure and must be taken into account in the analysis. This analysis is normally carried out using a substructure technique: the dynamic behaviour of the foundation (considered massless) together with the soil is studied first and then the foundation is connected to the rest of the structure and to its own mass.

The dynamic behaviour of foundations of systems subjected to external loads is determined by their stiffness (rigidity) matrix, that relates the force vector (forces and moments) applied to the foundation and the resulting displacement vector (displacements and rotations). Both forces and displacements are assumed to be time harmonic functions; being the foundation stiffness matrix a function of the excitation frequency ω. The foundation is usually massive and may be assumed to be rigid. Once the dynamic stiffness of the foundation is known, the response of the foundation including its mass, or of any structure supported on it, may be easily evaluated. The response to non-harmonic forces can be computed by Fourier transform techniques.

When the system is excited by waves travelling through the soil, prior to the analysis of the structure mounted on the springs defined by the foundation stiffness, the excitation of such system must be determined. To this end, the response of the massless surface or embedded foundation inpinged by waves travelling through the soil is computed (scattering problem). The existence of the foundation produces a filtering of the waves, being the foundation response a function of its own geometry and of the incident waves (kinematic interaction). The resulting motion is applied to the base of the soil-structure model in order to compute the response of the structure to the incoming waves.

During the first half of this century machine foundations were design by rules-of-thumb or very simple models: for instance, Winkler type models. A revision of those methods may be found in Refs. [1, 2]. The first study of the stiffness of a foundation representing the soil as a linear elastic half-space was carried out by Reissner in 1936 [3]. He studied the response of a disc on the surface of the soil subjected to vertical harmonic forces. A uniform distribution of stresses under the disc was assumed. Knowing that the actual stress distribution was far from being uniform, in the middle 1950s, several authors did studies assuming certain stress distributions for circular and rectangular foundations [4, 5, 6]. The mixed boundary value problem, with prescribed displacements under the rigid footing and zero tractions over the remaining portion of the surface, was studied during the 1960s and early 1970s [7, 8, 9, 10]. Relaxed boundary conditions were assumed under the footing. Several works were also done for viscoelastic soil models [11, 12].

Wong and Luco [13] computed dynamic compliances (stiffness inverse) of a surface rigid massless foundation of arbitrary shape on an elastic half-space by dividing the soil-foundation interface into rectangular elements. The tractions were considered uniformly distributed within the elements and a relation between the tractions over an element and the displacements on the soil surface was obtained by integration of Lamb's solution [14]. This method is in fact a Boundary Element Method with a fundamental solution for the half-space. However, the integration of the fundamental solution is rather involved and only surface foundations may be analyzed. Other authors presented similar approximate Methods [15, 16].

The Boundary Element Methods (BEM) are very well suited for the analysis of boundless regions and so they present advantages when used to compute foundation stiffnesses. The most extended formulation of the BEM (direct formulation and complete space fundamental solution) can be very easily applied to compute dynamic stiffnesses of surface or embedded foundations. The frequency domain BEM formulation was first used to obtain stiffnesses of rectangular foundations resting on, or embedded in, a viscoelastic half-space by Dominguez [17]. Non-homogeneous soils have been also studied by Abascal [18] and Abascal and Dominguez [19]; and circular foundations on layered soils by Gomez-Lera et al. [20]. Apsel in 1979 [21] used an indirect BEM in combination with semiexplicit Green's functions to compute stiffnesses of circular foundations embedded in a layered half-space. More recently, Karabalis and Beskos [22], and Karabalis, Spyrakos and Beskos [23] computed dynamic stiffnesses of massless surface foundations excited by non-harmonic forces using time domain BEM.

It should be said that during the last twenty years the most extended numerical technique for computation of foundation stiffnesses has been the Finite Element Method (FEM). The development of energy absorbing boundaries for 2-D [24] and axisymmetric problems [25] made possible the analysis of foundations resting on, or embedded in, layered soils. The finite elements models, however, present requirements like the necesity of a rigid bedrock at the bottom, or layers geometry that must be parallel and extending to infinity, that are not always realistic. In addition, 3-D dynamic analysis of foundations present important difficulties even though some approaches have been proposed to represent the 3-D behaviour. The BEM, on the contrary, may represent easily the boundless 3-D soil domain and do not present the aforementioned requirements.

The analysis of the response of foundations to incoming waves taking into account the interaction between the soil and the footing is a problem of diffraction of elastic waves. That kind of problems for inclusions and cavities have been treated by numerous authors since the early 1970s. Most of the existing exact solutions correspond to 2-D antiplane models [26, 27, 28, 29, 30]. An indirect BEM (the source method) was used to study the diffraction of SH, SV and P waves by canyons and alluvial valleys by Wong [31] and Dravinsky [32, 33], respectively. Wong and Luco [34] studied the response of 3-D surface foundations to travelling waves using the same boundary method they had used to compute foundations stiffnesses, and that has been mentioned above. Kobori, Minai and Shinozaki [35] and Luco [36] studied the torsional response of axisymmetric structures resting on the surface, to obliquely incident SH waves, using similar analytical procedures.

The BEM was applied to seismic waves diffraction problems by Dominguez [37]. He studied the response of 3-D surface and embedded foundations to incident SH, SV and P waves, assuming a homogeneous viscoelastic soil. Non-homogeneous soils have been considered for 2-D foundations by Abascal and Dominguez [38]. Spyrakos [39] and Karabalis [40] studied the response of foundations to travelling waves by means of the time domain BEM formulation, assuming a homogeneous elastic soil.

In the next two sections, the basic equations of the dynamic BEM are studied. The integral representations are obtained from the reciprocal theorem and the fundamental solution. It is done first for the general elastodynamic problem and then for steady-state. In the latter, the direct and indirect representations are shown.

The discretization of the integral representations to obtain the time domain and frequency domain BEM equations is summarized in Sects. 2.4 and 2.5.

In Sect. 2.6 the general foundation stiffness problem is presented; while, Sects. 2.7–2.9 are dedicated to the computation of stiffnesses for 3-D, 2-D and axisymmetric foundations, respectively.

Section 2.10 treats the response of foundations to travelling harmonic waves. Surface and embedded, strip and rectangular foundations are studied.

The chapter is closed by Sect. 2.11, where a revision of the use of time domain BEM for the dynamic analysis of foundations is done.

2.2 Integral Representation of the Displacements of the General Elastodynamic Problem

In this section the fundamental solution to the field equations corresponding to a concentrated point load is presented first. Then the reciprocal theorem is formulated and subsequently the integral representation for the displacements is obtained. In the following only the main formulae will be shown. A complete development may be found in Ref. [41] which follows [42].

2.2.1 Fundamental Solution

The response of an infinite elastic space to a time dependent point load is due to Stokes [43]. The load is a body force that may be written as

$$\rho f(x, t) = f(t)\delta(x - \xi)e, \tag{1}$$

where x stands for a general point of the space, ξ is the point where the concentrated load following e is applied, δ is the Dirac delta function, $f(t)$ represents the time variation of the force and ρ is the density. Using Helmholz's representation to decompose the force into its irotational and solenoidal parts, Navier's equation may be transformed into two wave equations which solutions lead by derivation to the displacements vector:

$$u_i = u_{ik}e_k, \tag{2}$$

where u_{ik} represents the displacement in the i direction when the point load is applied following k, and has the form:

$$
\begin{aligned}
u_{ik}(x, t; \xi | f) = \frac{1}{4\pi\rho} \Bigg\{ &\left(\frac{3r_i r_k}{r^3} - \frac{\delta_{ik}}{r} \right) \int_{C_p^{-1}}^{C_s^{-1}} \lambda f(t - \lambda r) \, d\lambda \\
&+ \frac{r_i r_k}{r^3} \left[\frac{1}{C_p^2} f\left(t - \frac{r}{C_p} \right) - \frac{1}{C_s^2} f\left(t - \frac{r}{C_s} \right) \right] \\
&+ \frac{\delta_{ik}}{rC_s^2} f\left(t - \frac{r}{C_s} \right) \Bigg\}
\end{aligned}
\tag{3}
$$

being $r = x - \xi$, $r = |r|$, C_p and C_s the P and S waves velocities, respecitvely, δ_{ij} the Kronecker's delta and λ a dummy variable.

The corresponding stresses may be obtained by means of Hooke's law and have the form:

$$\sigma_{ij} = \sigma_{ijk}e_k \tag{4}$$

being

$$
\begin{aligned}
\sigma_{ijk}(x, t; \xi | f) = \frac{1}{4\pi} \Bigg\{ &-6C_s^2 \left[5\frac{r_i r_j r_k}{r^5} - \frac{\delta_{ij}r_k + \delta_{ik}r_j + \delta_{jk}r_i}{r^3} \right] \int_{C_p^{-1}}^{C_s^{-1}} \lambda f(t - \lambda r) \, d\lambda \\
&+ 2\left[6\frac{r_i r_j r_k}{r^5} - \frac{\delta_{ij}r_k + \delta_{ik}r_j + \delta_{jk}r_i}{r^3} \right] \left[f\left(t - \frac{r}{C_s} \right) \right. \\
&\left. - \frac{C_s^2}{C_p^2} f\left(t - \frac{r}{C_p} \right) \right] + 2\frac{r_i r_j r_k}{r^4 C_s} \left[\dot{f}\left(t - \frac{r}{C_s} \right) - \frac{C_s^3}{C_p^3} \dot{f}\left(t - \frac{r}{C_p} \right) \right] \\
&- \frac{r_k \delta_{ij}}{r^3} \left(1 - 2\frac{C_s^2}{C_p^2} \right) \left[f\left(t - \frac{r}{C_p} \right) + \frac{r}{C_p} \dot{f}\left(t - \frac{r}{C_p} \right) \right] \\
&- \frac{\delta_{ik}r_j + \delta_{jk}r_i}{r^3} \left[f\left(t - \frac{r}{C_s} \right) + \frac{r}{C_s} \dot{f}\left(t - \frac{r}{C_s} \right) \right] \Bigg\}.
\end{aligned}
\tag{5}
$$

Tractions on a surface which outer normal is n are:

$$t_i^n = \sigma_{ij}n_j \quad \text{and} \quad t_{ik}^n = \sigma_{ijk}n_j. \tag{6}$$

Explicit expressions for u_{ik} and σ_{ijk} for the case that the body force time dependence is also a δ-function, i.e. $f(t) = \delta(t)$, can be easily obtained taking into account that

$$\int_{c_p^{-1}}^{c_s^{-1}} \lambda\delta(t - \lambda r)\,d\lambda = \frac{t}{r^2}\left[H\left(t - \frac{r}{C_p}\right) - H\left(t - \frac{r}{C_s}\right)\right]. \tag{7}$$

2.2.2 Reciprocal Theorem

This theorem is the dynamic counterpart of the classical Betti's reciprocal theorem of elastostatics. It was obtained by Graffi [44] in 1946–1947 and extended by Wheeler and Sternberg [42] to unbounded regions in 1978. Let Ω be a regular region with boundary Γ and consider two elastodynamic states with body forces displacements and tractions vectors f, u and t^n, respectively, for the first, and f', u' and t'^n for the second. The initial conditions are

$$u(x,0) = u_0(x); \quad u(x,0) = v_0(x)$$
$$u'(x,0) = u'_0(x); \quad \dot{u}'(x,0) = v'_0(x). \tag{8}$$

Then for $t \geqslant 0$

$$\int_{\Gamma} [t^n * u']\,d\Gamma + \rho \int_{\Omega} [f * u' + v_0 u' + u_0 \dot{u}']\,d\Omega$$
$$= \int_{\Gamma} [t'^n * u]\,d\Gamma + \rho \int_{\Omega} [f' * u + v'_0 u + u'_0 \dot{u}]\,d\Omega, \tag{9}$$

where the symbol $*$ stands for the convolution product. When both states have zero initial conditions (quiescent past), Eq. (9) becomes:

$$\int_{\Gamma} [t^n * u']\,d\Gamma + \rho \int_{\Omega} [f * u']\,d\Omega = \int_{\Gamma} [t'^n * u]\,d\Gamma = \rho \int_{\Omega} [f' * u]\,d\Omega. \tag{10}$$

2.2.3 Direct Integral Representation

The reciprocal theorem will be applied between the actual state and that corresponding to a unit concentrated impulse load following direction k at a point ξ of region Ω which is considered to be included in the infinite domain.

$$\rho f'_i = \delta(t)\delta(x - \xi)\delta_{ik}. \tag{11}$$

For the last state, zero initial conditions are prescribed

$$u'_0 = v'_0 = 0 \tag{12}$$

while displacements and tractions may be written as

$$u'_i = U_{ik}(x, t; \xi),$$
$$t'^n_i = T^n_{ik}(x, t; \xi), \tag{13}$$

where U_{ik} and T^n_{ik} are equal to u_{ik} and $t^n_{ik} = \sigma_{ijk}n_j$, given by Eqs. (3) and (5), when the body force is a unit concentrated pulse.

Equation (9) transforms into

$$u_k(\xi, t) = \int_{\Gamma} [U_{ik} * t^n_i - T^n_{ik} * u_i]\,d\Gamma$$
$$+ \rho \int_{\Omega} [U_{ik} * f_i + U_{ik} * v_{0_i} + \dot{U}_{ik}u_{0_i}]\,d\Omega. \tag{14}$$

It can be easily shown that the convolution products may be written as

$$U_{ik} * t_i^n = u_{ik}[x, t; \xi | t_i^n(x, t)]$$
$$U_{ik} * f_i = u_{ik}[x, t; \xi | f_i(x, t)]$$

(15)

$$T_{ik}^n * u_i = t_{ik}^n[x, t; \xi | u_i(x, t)]$$

(16)

and Eq. (14) transforms into

$$u_k(\xi, t) = \int_\Gamma \{u_{ik}[x, t; \xi | t_i^n(x, t)] - t_{ik}^n[x, t; \xi | u_i(x, t)]\} d\Gamma$$

$$+ \rho \int_\Omega \{u_{ik}[x, t; \xi | f_i(x, t)] + U_{ik} v_{0_i} + \dot{U}_{ik} u_{0_i}\} d\Omega.$$

(17)

Equation (17) is known as Love's [45] integral identity and it gives the displacement in the direction k at any interior point ξ of region Ω as a summation of three parts. One is a boundary integral of the displacements and tractions corresponding to the fundamental solution for values of the point force at ξ, equal to the actual problem tractions and displacements, respectively, at the integration point. The second is the integral through Ω of the displacements of the fundamental solution for values of the point force at ξ equal to the actual body force at the integration point. The third is the integral through Ω of the products of initial conditions and displacements/velocities of the fundamental solution for a unit concentrated pulse.

The integral representation is also valid for boundary and external points if it is written in the form:

$$C(\xi)u_k(\xi, t) = \int_\Gamma \{u_{ik}[x, t; \xi | t_i^n(x, t)] - t_{ik}^n[x, t; \xi | u_i(x, t)]\} d\Gamma$$

$$+ \rho \int_\Omega \{u_{ik}[x, t; \xi | f_i(x, t)] + U_{ik} v_{0_i} + \dot{U}_{ik} u_{0_i}\} d\Omega,$$

(18)

where

$$C(\xi) = \begin{cases} 1 & \text{if } \xi \in \Omega \\ \frac{1}{2} & \text{if } \xi \in \Gamma \text{ being } \Gamma \text{ smooth at } \xi \\ 0 & \text{if } \xi \in \Omega_c \end{cases}$$

(19)

being Ω_c the complement of Ω in the infinite region.

2.3 Integral Representations of the Displacements of the Steady-state Problem

All the formulae shown in the previous section simplify when the steady-state situation is considered. The body forces and boundary conditions have a time variation of the type $\exp(i\omega t)$. This behaviour is assumed from $t = -\infty$ and consequently there is not transient part in the solution. All variables of the response are also harmonic in time with an angular frequency ω.

The study of steady-state problems is the base for the frequency domain analysis of linear elastodynamic problems using the Fourier integral transform.

2.3.1 Fundamental Solution

This solution corresponds to a point load at ξ which time variation is $\exp(i\omega t)$. Substituting $f(t) = \exp(i\omega t)$ in Eq. (3), the following expression is obtained

$$U_{ik}(x,\xi,\omega) = u_{ik}(x,t;\xi \exp(i\omega t)) = \frac{1}{\alpha\pi\rho C_s^2}[\psi\delta_{ik} - \chi r_{,i}r_{,j}], \tag{20}$$

where $\alpha = 4$,

$$\psi = \left(1 - \frac{C_s^2}{\omega^2 r^2} + \frac{C_s}{i\omega r} + 1\right)\frac{\exp(-i\omega r/C_s)}{r}$$
$$- \left(\frac{C_s^2}{C_p^2}\right)\left(-\frac{C_p^2}{\omega^2 r^2} + \frac{C_p}{i\omega r}\right)\frac{\exp(-i\omega r/C_p)}{r} \tag{21}$$

and

$$\chi = \left(-\frac{3C_s^2}{\omega^2 r^2} + \frac{3C_s}{i\omega r} + 1\right)\frac{\exp(-i\omega r/C_s)}{r}$$
$$- \left(\frac{C_s^2}{C_p^2}\right)\left(-\frac{3C_p^2}{\omega^2 r^2} + \frac{3C_p}{i\omega r} + 1\right)\frac{\exp(-i\omega r/C_p)}{r}. \tag{22}$$

Tractions on a surface which outer normal is n are obtained from Eqs. (5) and (6).

$$T_{ik}^n(x,\xi,\omega) = t_{ik}^n(x,t;\xi|\exp(i\omega t))$$
$$= \frac{1}{\alpha\pi}\left\{\left(\frac{d\psi}{dr} - \frac{1}{r}\chi\right)\left(\delta_{ij}\frac{\partial r}{\partial n} + r_{,i}n_j\right) - \frac{2}{r}\chi\left(n_i r_{,j} - 2r_{,i}r_{,j}\frac{\partial r}{\partial n}\right)\right.$$
$$\left. - 2\frac{d\chi}{dr}r_{,i}r_{,j}\frac{r}{n} + \left(\frac{C_p^2}{C_s^2} - 2\right)\left(\frac{d\psi}{dr} - \frac{d\chi}{dr} - \frac{\alpha}{2r}\chi\right)r_{,j}n_i\right\}. \tag{23}$$

The steady state fundamental solution can also be easily obtained from Navier's equation particularized for steady-state problems. In Ref. [46] that derivation may be found and also a study of the convergency of both the 3-D and 2-D solutions to the static ones when the frequency tends to zero. That analysis is particularly interesting for two dimensions, since an undetermined constant appears in the static case but not in dynamics.

The steady-state fundamental solution for two dimensions is also given by Eqs. (20) and (23) being $\alpha = 2$,

$$\psi = K_0\left(\frac{i\omega r}{C_s}\right) + \frac{C_s}{i\omega r}\left[K_1\left(\frac{i\omega r}{C_s} - \frac{C_s}{C_p}K_1\frac{i\omega r}{C_p}\right)\right] \tag{24}$$

and

$$\chi = K_2\left(\frac{i\omega r}{C_s}\right) - \frac{C_s^2}{C_p^2}K_2\left(\frac{i\omega r}{C_p}\right). \tag{25}$$

Functions K_0, K_1 and K_2 are the modified Bessel functions of the second kind and order 0, 1 and 2, respectively.

2.3.2 Reciprocal Theorem

The reciprocal theorem given by Eq. (9) becomes more simple when both states are assumed to be harmonic with the same frequency. The convolution products transform into normal products and Eq. (9) gives

$$\int_{\Gamma} t'' u' \, d\Gamma + \rho \int_{\Omega} f u' \, d\Omega = \int_{\Gamma} t'' u \, d\Gamma + \rho \int_{\Omega} f' u \, d\Omega, \tag{26}$$

where all variables represent amplitudes and the $\exp(i\omega t)$ terms are implicit.

2.3.3 Direct Integral Representation

The reciprocal theorem is written for the actual steady-state problem and another one corresponding to a unit concentrated harmonic load that follows direction k and has the same frequency of the actual state.

$$\rho f_i' = \delta(x - \xi)\delta_{ik}. \tag{27}$$

Equation (26) transforms into

$$C(\xi)u_k(\xi, \omega) = \int_{\Gamma} (U_{ik} t_i^n - T_{ik}^n u_i) \, d\Gamma + \rho \int_{\Omega} U_{ik} f_i \, d\Omega, \tag{28}$$

where $C(\xi)$ has the same meaning given by Eq. (19) for the general problem.

2.3.4 Indirect Integral Representation

Displacements at any internal point of region Ω, when body forces are zero, may be represented in terms of the single-layer potentials $U_{ik}(x, \xi, \omega)$ (see, for instance, Ref. [47])

$$u_k(\xi, \omega) = \int_{\Gamma} s_i(x) U_{ik}(x, \xi, \omega) \, d\Gamma, \tag{29}$$

where $s_i(x, \omega)$ represents the density of the potential at boundary points. A similar representation in terms of the double-layer potential $T_{ik}^n(x, \xi, \omega)$ may be written.

$$u_k(\xi, \omega) = \int_{\Gamma} d_i(x) T_{ik}^n(x, \xi, \omega) \, d\Gamma. \tag{30}$$

The above integral representations are known as indirect representations because u is given in terms of densities s_i or d_i which are neither tractions nor displacements over the boundary as they were in the direct representation. However, $s_i(d_i)$ may be interpreted as the jump of tractions (displacements) between the internal and the external problems when both have the same boundary displacements (tractions).

A useful representation of the tractions at ξ on a surface which normal is λ may be easily obtained from Eq. (29). Since $U_{ik}(x, \xi, \omega) = U_{ik}(\xi, x, \omega)$, it may be written

$$u_k(\xi, \omega) = \int_{\Gamma} s_i(x) U_{ik}(\xi, x, \omega) \, d\Gamma \tag{31}$$

and by derivation at ξ

$$t_k^{\lambda}(\xi, \omega) = \int_{\Gamma} s_i(x) T_{ik}^{\lambda}(\xi, x, \omega) \, d\Gamma. \tag{32}$$

Equation (29) remains the same when ξ is at boundary Γ. On the other hand, according to the properties of the double-layer potentials, Eqs. (30) and (32) transform into integral equations of the second kind.

$$u_k(\xi, \omega) = -\tfrac{1}{2}\delta_{ik}d_i(\xi) + \int_\Gamma d_i(x)T_{ik}^n(x, \xi, \omega)\,d\Gamma \qquad (33)$$

and

$$t_k^\lambda(\xi, \omega) = -\tfrac{1}{2}\delta_{ik}s_i(\xi) + \int_\Gamma s_i(x)T_{ik}^\lambda(\xi, x, \omega)\,d\Gamma. \qquad (34)$$

2.4 Time Domain Boundary Element Method

Applying boundary conditions to the integral representation [Eq. (18)], written for boundary points, one obtains a boundary integral equation that may be solved numerically. To do so, the time interval $0 \to t$ is divided into time steps of length Δt. Displacements and tractions are assumed to be constant along each time step (Fig. 1). The boundary of the domain is assumed to consist of a number of plane elements being displacements and tractions constant over each boundary element (Fig. 2) and their values assigned to a nodal point.

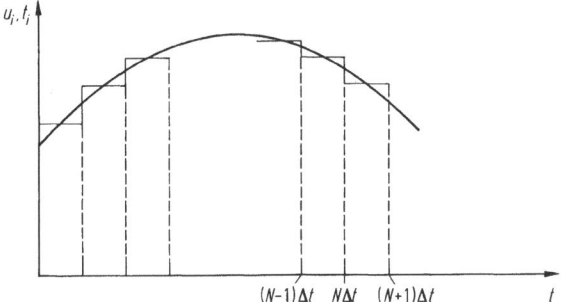

Fig. 1. Time discretization

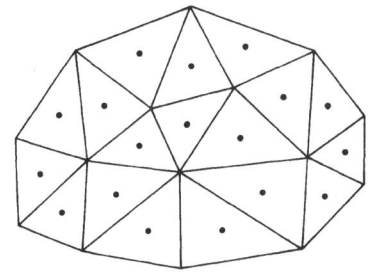

Fig. 2. Constant boundary elements for 3-D problems

Assuming zero body forces and initial conditions, Eq. (18) may be written for a boundary point ξ as

$$\tfrac{1}{2}u_k(\xi, t) = \int_\Gamma \{u_{ik}[x, t; \xi|t_i^n(x, t)] - t_{ik}^n[x, t; \xi|u_i(x, t)]\}\, d\Gamma \tag{35}$$

being Γ smooth at ξ.

In order to accomplish the time stepping process, displacements and tractions are written for time interval l as

$$u_i^l(x, t) = u_i^l(x)[H(t - (l - 1)\Delta t) - H(t - l\Delta t)]$$
$$t_i^{nl}(x, t) = t_i^{nl}(x)[H(t - (l - 1)\Delta t) - H(t - l\Delta t)]. \tag{36}$$

The kernels of Eq. (35), when tractions and displacements corresponding to the first time steps are applied, become

$$u_{ik}(x, t; \xi|t_i^{n1}(x, t)) = \frac{1}{4\pi\rho}\left\{\frac{1}{r^2}\left(\frac{3r_i r_k}{r^3} - \frac{\delta_{ik}}{r}\right)(F_p - F_s)t_i^{n1}(x) + \frac{1}{C_p^2}\frac{r_i r_k}{r^3}t_i^1\left(x, t - \frac{r}{C_p}\right)\right.$$
$$\left. + \frac{1}{C_s^2}\left(\frac{\delta_{ik}}{r} - \frac{r_i r_k}{r^3}\right)t_i^1\left(x, t - \frac{r}{C_s}\right)\right\} \tag{37}$$

and

$$t_{ik}^n(x, t; \xi|u_i^1(x, t)) = \sigma_{ijk}(x, t; \xi|u_i^1(x, t))n_j$$

$$\sigma_{ijk}(x, t; \xi|u_i^1(x, t)) = \frac{1}{4\pi}\left(-6C_s^2\left(5\frac{r_i r_j r_k}{r^7} - \frac{\delta_{ij}r_k + \delta_{ik}r_j + \delta_{jk}r_i}{r^5}\right)(F_p - F_s)u_i^1(x)\right.$$

$$- \left[2\frac{C_s^2}{C_p^2}\left(6\frac{r_i r_j r_k}{r^5} - \frac{\delta_{ij}r_k + \delta_{ik}r_j + \delta_{jk}r_i}{r^3}\right) + \left(1 - 2\frac{C_s^2}{C_p^2}\right)\frac{r_k\delta_{ij}}{r^3}\right]$$

$$\cdot u_i^1\left(x, t - \frac{r}{C_p}\right) + \left[2\left(6\frac{r_i r_j r_k}{r^5} - \frac{\delta_{ij}r_k + \delta_{ik}r_j + \delta_{jk}r_i}{r^3}\right)\right.$$

$$\left.\left. - \frac{\delta_{ik}r_j + \delta_{jk}r_i}{r^3}\right]u_i^1\left(x, t - \frac{r}{C_s}\right)\right\}, \tag{38}$$

where

$$F_\alpha = \begin{cases} 0 & \text{when } t - (r/C_\alpha) \leqslant 0 \\ \tfrac{1}{2}[t^2 - (r/C_\alpha)^2] & \text{when } 0 < t - (r/C_\alpha) < \Delta t \\ \tfrac{1}{2}[t^2 - (t - t)^2] & \text{when } t - (r/C) > \Delta t \end{cases} \tag{39}$$

and α stands for p or s.

Using Eqs. (35), (37) and (38), the displacement at a point ξ when $t = N\Delta t$ may be written in terms of boundary tractions and displacements, at the same and all previous time steps.

$$\tfrac{1}{2}u_k^N = \sum_{n=1}^N \int_\Gamma \{u_{ik}[x, t; \xi|t_i^{nl}(x, t)] - t_{ik}^n[x, t; \xi|u_i^l(x, t)]\}\, d\Gamma, \tag{40}$$

where $l = N - n + 1$.

The integrals of Eq. (40) are non-zero for a part of the boundary that depends on the value of n and the kind of wave that each quantity represents. For instance:

$$\int_{\Gamma} u_{ik}[x, t; \xi | t_i^{nl}(x, t)] \, d\Gamma = \int_{\Gamma_c} \left(\frac{3r_i r_k}{r^5} - \frac{\delta_{ik}}{r^3}\right)(F_p - F_s) t_i^{nl}(x) \, d\Gamma$$

$$+ \frac{1}{C_p^2} \int_{\Gamma_{c_p}} \frac{r_i r_k}{r^3} t_i^{nl}(x) \, dB$$

$$+ \frac{1}{C_s^2} \int_{\Gamma_{c_s}} \left(\frac{\delta_{ik}}{r} - \frac{r_i r_k}{r^3}\right) t_i^{nl}(x) \, dB, \qquad (41)$$

where Γ_c is the part of Γ where $C_s(n-1)\Delta t < r \leqslant C_p n\Delta t$,
Γ_{cp} is the part of Γ where $C_p(n-1)\Delta t < r \leqslant C_p n\Delta t$,
Γ_{cs} is the part of Γ where $C_s(n-1)\Delta t < r \leqslant C_s n\Delta t$.

A similar expression may be written for the second part of the integral in Eq. (40).

Discretizing the boundary into Q plane elements, Eq. (40), for the nodal point of the boundary element p, may be written as

$$\tfrac{1}{2} u_p^N = \sum_{n=1}^{N} \sum_{q=1}^{Q} (G_q^n t_q^l - H_q^n u_q^l), \qquad (42)$$

where $l = N - n + 1$, u_q^l and t_q^l are the displacements and tractions at node q and time step l,

$$G_{ikq}^n = \frac{1}{4\pi} \int_{\Delta\Gamma_q} \frac{1}{r^2} \left(\frac{3r_i r_k}{r^3} - \frac{\delta_{ik}}{r}\right)(F_p - F_s) \, d\Gamma$$

$$+ \frac{1}{C_p^2} \int_{\Delta\Gamma_q} \frac{r_i r_k}{r^3} \, d\Gamma + \frac{1}{C_s^2} \int_{\Delta\Gamma_q} \left(\frac{\delta_{ik}}{r} - \frac{r_i r_k}{r^3}\right) d\Gamma, \qquad (43)$$

where $\Delta\Gamma_q$ stands for the element q and the integrals are non-zero only for those elements that belong to Γ_C, Γ_{C_p}, Γ_{C_s} respectively. Again a similar but more lengthy expression may be written for H_{ikq}^n. G_q^n and H_q^n represent, respectively, the displacement and traction vectors at element q and time step n when a unit rectangular pulse is applied at p during the first time step.

Equation (42) may be written for every boundary node, and proceeding step by step the solution to the boundary integral equation, from $t = 0$ to $t = N\Delta t$, is obtained. Details about some of the numerical aspects of the time domain BEM may be found in Refs. [48, 40, 22].

2.4.1 Two-Dimensional Problems

For two dimensional problems the above formulation remains basically the same. Assuming a plane strain problem, the fundamental solution corresponds now to a load distributed along the x_3 axis. The boundary elements are straight segments and the nodes are located at the mid-point of each element (Fig. 3).

Equation (40) remains the same but being now the kernels u_{ik} and t_{ik} those obtained by integration of the three dimensional solution along the x_3-axis. It is

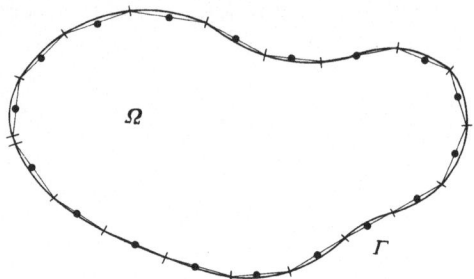

Fig. 3. Constant boundary elements for 2-D problems

obvious that for a certain time t, the integrals along x_3 only extend to that part of the axis that is at a distance from x smaller than the length travelled by the waves from 0 to t.

The counterpart of Eq. (37) for two-dimensions is

$$u_{ik}(x,t;\xi|t_i^{n1}(x,t)) = \frac{1}{2\pi\rho}\left(\frac{2}{\xi_i\xi_k}\left[H\left(t-\frac{r}{C_p}\right)\int_r^{C_p t}\frac{F_p\,d\eta}{(\eta^2-r^2)^{1/2}}\right.\right.$$
$$\left.-H\left(t-\frac{r}{C_s}\right)\int_r^{C_s t}\frac{F_s\,d\eta}{(\eta^2-r^2)^{1/2}}\right]t_i^{n1}(x)$$
$$\left.+\frac{\delta_{ik}}{C_s^2}-H\left(t-\frac{r}{C_s}\right)\int_r^{C_s t}\frac{t_i^n(x,t-C_s^\eta)}{(\eta^2-r^2)^{1/2}}\,d\eta\right\}, \qquad (44)$$

where

$$F_\alpha = \begin{cases} 0 & \text{when } t-(\eta/C_\alpha)<0 \\ \frac{1}{2}[t-(\eta/C_\alpha)]^2 & \text{when } 0<t-(\eta/C_\alpha)\leqslant\Delta t \\ [t-(\eta/C_\alpha)\Delta t-\frac{1}{2}\Delta t^2 & \text{when } t-(\eta/C_\alpha)>\Delta t \end{cases} \qquad (45)$$

and α stands for p or s.

Again, the problem is solved proceeding step by step to the solution of the system that is obtained writing Eq. (42) for every boundary node.

2.5 Frequency Domain Boundary Element Method

2.5.1 Direct Formulation

The integral representation [Eq. (28)] is particularized for zero body forces, and assuming that the boundary Γ is smooth at ξ one may write

$$\tfrac{1}{2}u_k(\xi,\omega) = \int_\Gamma U_{ik}t_i^n\,d\Gamma - \int_\Gamma T_{ik}^n u_i\,d\Gamma. \qquad (46)$$

Applying boundary conditions to Eq. (46) an integral equation is obtained. In order to solve the integral equation, the boundary is discretized into plane elements (three dimensions) as shown in Fig. 2 or straight segments (two dimensions) as shown in Fig. 3. Displacements and tractions are assumed to be constant over each element. Equation (46) may be written for every boundary node as

$$\tfrac{1}{2}u_p = \sum_{q=1}^{Q} [G_{pq}t_q - H_{pq}u_q] \quad \text{para } p = 1, 2, \ldots, \tag{47}$$

where G_{pq} and H_{pq} are square matrices which elements are integrals over the element q of U_{ik} and T_{ik}^n when the harmonic point load is applied at p.

For any boundary node and for each direction, either t or u is known; consequently, Eq. (47) provides as many equations as unknowns.

Needless to say that linear, quadratic and higher order elements may be used to discretize Eq. (46) (see, for instance, Refs. [49, 50]). The same can be said for the time domain formulation presented above.

2.5.2 Indirect Formulation

Equations (29), (33) or (34) may be used to represent displacements or tractions at boundary points in terms of the potential density over the boundary [51]. An integral equation is obtained and its solution computed using the same kind of discretization of the direct method but with the potential densities being now the unknowns.

One possibility of the indirect methods is to locate the sources of potential along a surface away from the boundary. This procedure simplifies the integration process but makes necessary to determine the location of the sources surface. Examples of its use for dynamic of foundations problems may be found in the work of Apsel [21], Wong [31] and Dravinsky [32, 33].

2.6 Dynamic Stiffness of Foundations

One of the main steps in the dynamic analysis of structures taking into account the soil flexibility is the determination of the foundation dynamic stiffness, also known as dynamic impedance.

In most problems where the soil-structure interaction effect is important (machine foundations, power plants), the foundation is massive and may be considered a rigid body. The foundation has three degrees of freedom corresponding to horizontal, vertical and rocking (rotation) coordinates when it is a strip footing that may be studied using a plane model (Fig. 4). For 3-D models of foundations

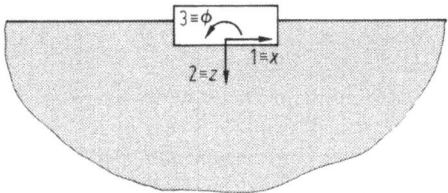

Fig. 4. Strip foundation

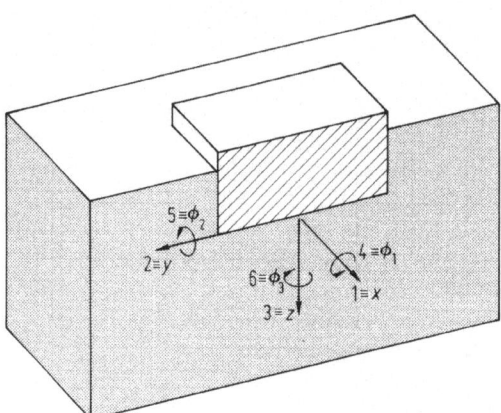

Fig. 5. Three-dimensional foundation

(Fig. 5) each vector has six components: one vertical, two horizontal, two rocking and one torsional. The exitation is assumed to be harmonic being able to consider other exitations by means of Fourier transform techniques.

For a particular harmonic exitation with frequency ω, the dynamic stiffness matrix is defined as the matrix that relates the vector of forces (and moments) applied to the foundation and the resulting vector of displacements (and rotations) when the foundation is considered massless:

$$R(t) = Ku(t). \tag{48}$$

The terms of the matrix K are functions of frequency ω and each term K_{ij} represents the resultant force or moment following coordinate i, of the contact tractions against the foundation when a unit harmonic displacement or rotation following coordinate j is applied to the foundation.—Properly speaking, the matrix K should be called stiffness or impedance of the soil for a given shape of the foundation.

It is worth to say that dynamic forces and displacements related by Eq. (48) are generally out of phase. It is convenient, then, to use complex notation to represent forces and displacements. The stiffnesses are also written as

$$K_{ij}(\omega) = \mathrm{Re}(K_{ij}) + \mathrm{i}\,\mathrm{Im}(K_{ij}), \tag{49}$$

where $\mathrm{i} = \sqrt{-1}$.

The real component of the stiffnesses reflects the stiffness and inertia of the soil. The imaginary component reflects the damping of the system. The main part of the damping is due to the energy dissipated by he waves propagating away from the foundation (radiation damping). It is obvious that since this kind of damping is due to the wave radiation, it exists for linear elastic half-space models or any other model that permits radiation of the waves. In addition to the radiation damping a histeretic material damping may exist.

The radiation damping is highly frequency dependent and for linear elastic soil models the stiffness components are usually written as

$$K_{ij} = k_{ij} + i a_0 c_{ij},$$ (50)

where k_{ij} and c_{ij} are frequency-dependent coefficients,
$a_0 = \omega B / C_s$ is a dimensionless frequency,
B is a characteristic length of the foundation, and
C_s is, as usual, the shear wave velocity.

When material damping exist an attempt to isolate the effect of that damping is done by writing the dynamic stiffness in the following form:

$$K_{ij} = (k_{ij} + i a_0 c_{ij})(1 + 2iD)$$ (51)

where D is the damping ratio. The coefficients k_{ij} and c_{ij} still depend on the material damping; however, for deep soil deposits and typical values of D, this dependence is small.

In a similar way to that used to define the stiffness matrix, one may define its inverse by

$$u(t) = F R(t).$$ (52)

The matrix F is known as dynamic compliance matrix or dynamic flexibility matrix and it is frequently used instead of the stiffness matrix. In the following, the stiffness matrix and the compliance matrix will used indistinctly. Following the complex notation above, the dynamic compliances may be written as

$$F_{ij}(\omega) = \mathrm{Re}(F_{ij}) + i \, \mathrm{Im}(F_{ij}).$$ (53)

2.7 Three-Dimensional Foundations

As was said before, the BEM are well suited for 3-D dynamic analysis of foundations since they can represent in a simple manner the half-space under the footing.

Using the frequency domain direct BEM formulation, dynamic stiffnesses of surface and embedded, square are rectangular, foundations are computed.

Due to the use of a fundamental solution corresponding to the complete space, not only the soil-foundation interface but also the soil free-surface should be discretized. However, in practice only a small region arround the foundation has to be included in the model since there is a small effect of the free-surface far away from the foundation on the computed values of the stiffness coefficients.

For all the problems analyzed, rectangular boundary elements are used. Tractions and displacements are constant over each element associated with a nodal point at its centre. Figure 6 shows one quarter of the boundary element discretization used for an embedded square foundation.

One column of the foundation stiffness matrix is obtained prescribing for each element under the footing the displacements that the nodal point would have as result of a unit rigid body motion of the foundation following a certain coordinate. Zero tractions are prescribed over the soil free-surface.

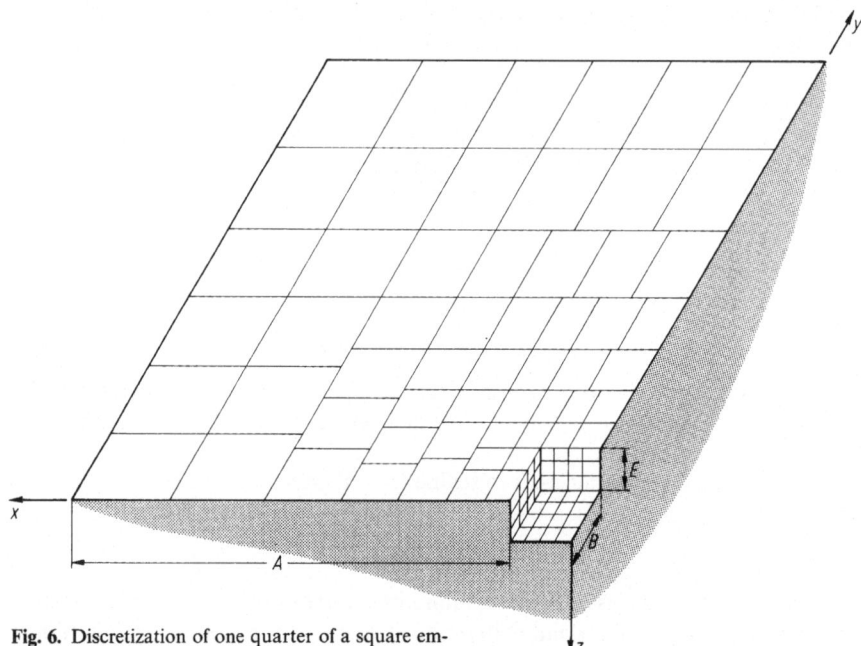

Fig. 6. Discretization of one quarter of a square embedded foundation

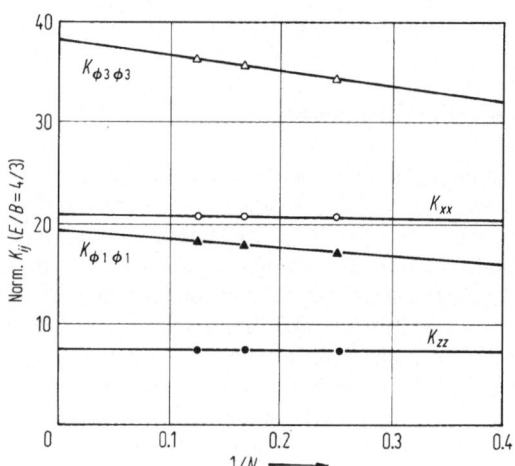

Fig. 7. Effect of the interface discretization on the static stiffnesses

2.7.1 Effect of the Number of Elements Under the Foundation

Even though the stress distribution under the footing has sharp peaks the BE mesh for the soil-foundation interface does not have to be very dense [17, 18] since the stress resultants over the foundation and not the stress distributions are needed.

In order to evaluate the effect of the size of the elements under the footing, stiffnesses of foundations with different levels of embedment and several frequencies may be computed. Figure 7 shows the variation of the computed static stiffnesses with the number of elements along half of the side of the bottom of a square embedded foundation ($E/B = 4/3$). The variation of the real and imaginary parts of the dynamic stiffnesses for a level of embedment $E/B = 2/3$ and a moderately high dimensionless frequency ($a_0 = 2.1$), is shown in Fig. 8. The stiffness components shown in Figs. 7 and 8 are normalized by the following factors: Norm. $K_{xx} = K_{xx}(2 - v)/GB$; Norm. $K_{zz} = K_{zz}(1 - v)/GB$; Norm. $K_{\phi2\phi2} = K_{\phi2\phi2}(1 - v)/GB^3$; and Norm. $K_{\phi3\phi3} = K_{\phi3\phi3}/GB^3$.

As may be seen in Figs. 7 and 8, the variation of all the stiffness functions with $1/N$ is approximately linear. Values corresponding to infinitely small elements ($1/N = 0$) can be extrapolated. Similar variation of the stiffnesses with $1/N$ are observed for other frequencies and embedment levels. An approach that would imply the solution of any problem using several discretizations and an extrapolation for $1/N = 0$ can be considered unnecessary because of the small values of the tangent shown by the lines in Figs. 7 and 8. The values computed with $N \geqslant 6$ are accurate enough if the condition of having at least 6 elements per wave length is also satisfied.

The study of the effect of the amount of soil free-surface included in the BE model that is presented below, has been done with $N = 8$; i.e. 64 elements in one quarter of the bottom of the footing, and a similar size for the elements of the walls when the foundation is embedded. Thus, the errors due to the discretization of the soil-foundation interface may be considered neglectible.

2.7.2 Effect of the Extension of the Soil Free-surface Discretization

The study of the influence of the amount of soil free-surface arround the foundation included in the model has been carried out using the surface discretization shown in Fig. 6 and all those that may be obtained from it by successive deletion of the outer most line of elements. The maximun distance from the footing reached by the model is $A = 7.5\ B$. The maximun length of one side of the elements used on the soil free-surface is equal to $\lambda/2 = \pi B/a_0$ for the elements more distant from the foundation.

Stiffnesses of square foundations with several different levels of embedment have been computed. In the following some results for a surface foundation, a foundation with moderate embedment ($E/B = 2/3$) and another deeply embedded ($E/B = 2$) are shown. In all cases the soil is assumed to be linear elastic with density $\rho = 1$ and Poisson's modulus $v = 1/3$.

Figure 9 shows the values of the real and imaginary parts of the horizontal stiffness versus the distance A from the footing that is reached by the discretization. The coefficients are normalized by GB and $(2 - v)$ which is known to be the

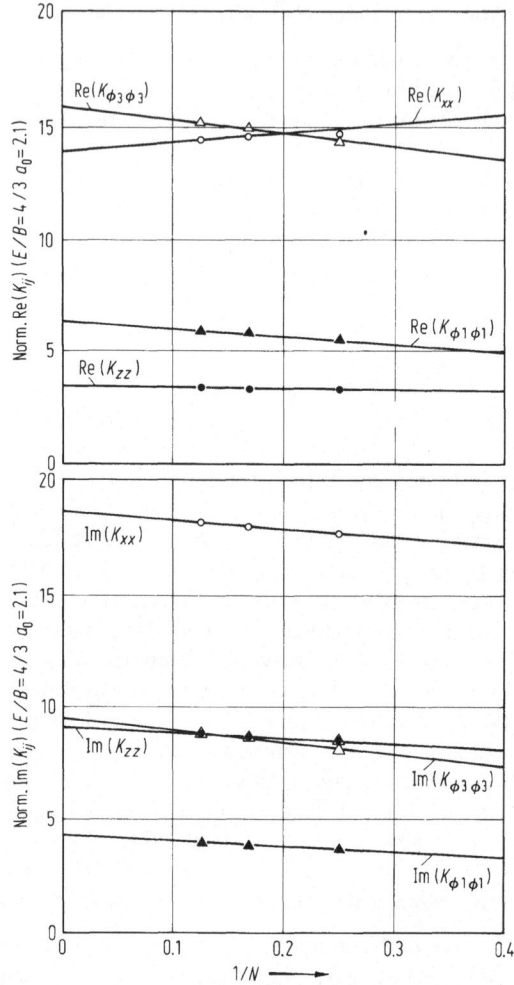

Fig. 8. Effect of the interface discretization on the dynamic stiffnesses

approximate Poisson's modulus dependence of K_{xx}. Results are plotted for four different values of the dimensionless frequency $a_0 = \omega B / C_s$.

It is apparent that the distance A has almost no effect on the computed values of k_{xx} and c_{xx}. The same happens for the vertical, rocking and torsional stiffness components (see Ref. [52]). Consequently, for surface foundations, and at least for values of $a_0 \leqslant 2.1$, all the stiffnesses can be computed without any discretization of the soil free-surface.

The study of the effect of the distance A on the stiffness of embedded foundations is done for two different fundamental solutions. One is the usual Kelvin type fundamental solution (single solution); and the other, consist of two point loads:

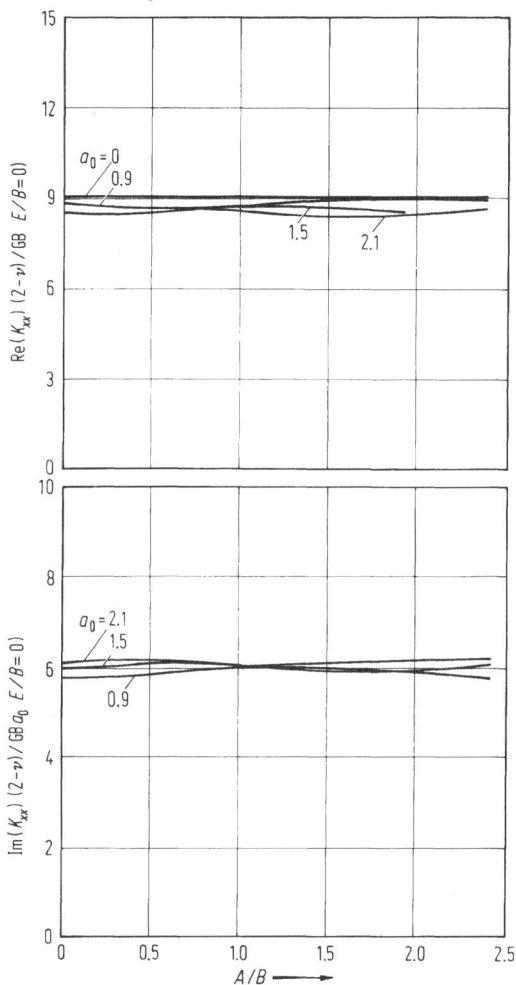

Fig. 9. Horizontal stiffness coefficients, Surface foundation

one located at the reference point and the other, having the same direction and sense, at the image of the first point with respect to the soil free-surface (double solution). The use of the latter makes tensor T, for points of the soil free-surface, to have all the terms equal to zero except T_{13}, T_{23}, T_{31} and T_{32}, being the integrals of these terms times the corresponding surface displacements the only quantities neglected over the non-discretized soil surface.

The values of the swaying and rocking coefficients for a foundation with an embedment $E/B = 2/3$ are shown, versus the distance A, in Fig. 10. Again, four different values of a_0 are considered and results computed using the two different fundamental solutions are now represented. It may be seen in Fig. 10 that the

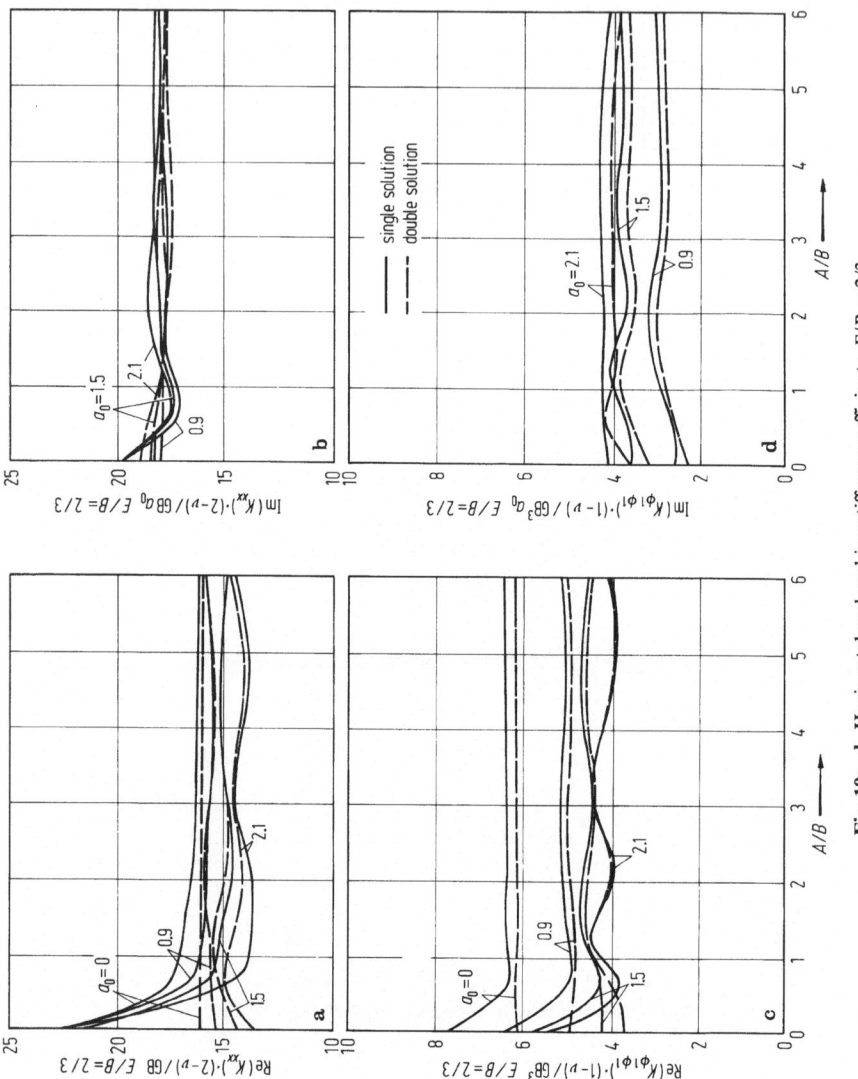

Fig. 10a–d. Horizontal and rocking stiffness coefficients. $E/B = 2/3$

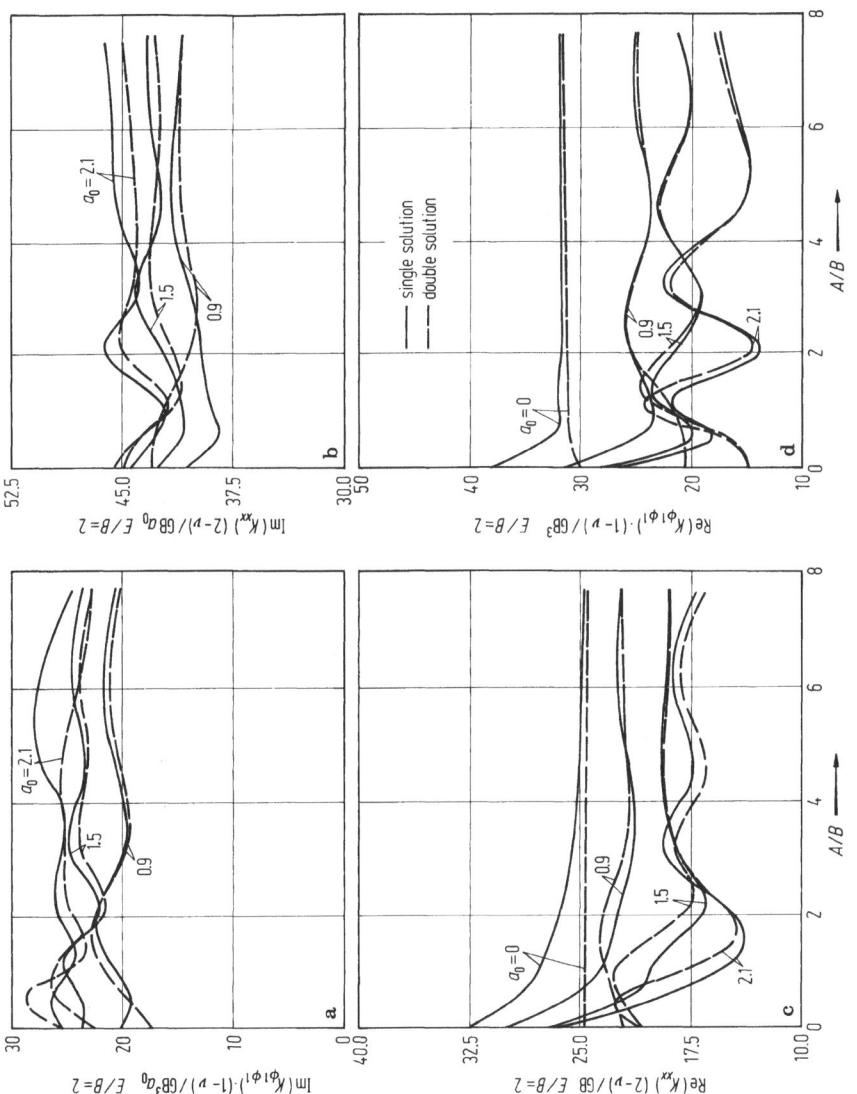

Fig. 11a–d. Horizontal and rocking stiffness coefficients. $E/B = 2$

computed values present certain oscillations when the amount of soil free-surface included in the model varies. The oscillations grow with a_0 but, in any case, when $a_0 \leqslant 2.1$ and $A/B \geqslant 2.5$ these oscillations are very small and the effect of the non-discretized free-surface can be neglected. The use of the double fundamental solution improves the results for low values of A/B but that improvement disappears when A/B grows. This behaviour may be explained by the fact that the terms of T that are non-zero in the double solution damp out much slower than the others for increasing values of $\omega r/C_s$ (r: distance to the load). Thus, for values of A and a_0 bigger than a certain amount, the remaining terms lose their relative importance also for the single solution. The vertical stiffness shows a behaviour versus A/B completely similar to that of the horizontal, being the torsion coefficients almost insensible to A/B [51].

The horizontal and rocking stiffness coefficients for a deeply embedded foundation ($E/B = 2$) are shown in Fig. 11 versus A/B. The effect of the amount of soil free-surface included in the model is important up to distances A bigger than those of the previous case, and increases rapidly with frequency. In general, it may be said that when $a_0 \leqslant 2.1$ values of $A/B \geqslant 7.5$ are big enough to take into account the effect of the soil free-surface. The vertical stiffness coefficients show a behaviour similar to the horizontal, being the effect of A very small for the torsion coefficients (see Ref. 52).

According to the above analysis of the effects of the amount of soil free-surface included in the model, it may be suggested that for square surface or embedded foundations and $a_0 \leqslant 2.1$, the soil free-surface around the foundation be discretized ap to a distance from the footing equal to four times the foundation embedment ($A = 4E$). The use of the double solution may be very useful to compute approximate values of the stiffnesses by discretization only of the soil foundation interface. The approximation gets better as the dimensionless frequency decreases and is very good for the static values.

2.7.3 Static Stiffnesses

As an example of the use of the BEM for computation of stiffnesses of 3-D foundations, the static stiffnesses of square foundations with several levels of embedment have been computed. The soil is assumed to be linear elastic and the double fundamental solution is used. Only the soil foundation interface has been discretized.

Figure 12 shows the variation of the normalized values of the static stiffness coefficients of square foundations versus the level of embedment. Since other results for these stiffnesses are not available for comparison, the computed values are compared with those obtained by Apsel [21] using a indirect BEM for the inscript and circumscript cylindrical foundations. A comparison is also done with values proposed for cylindrical foundations by Elsabee and Morray [53] and Kausel and Ushijima [54], and that are based on finite element models. As can be seen from Fig. 12, the BEM values are always between those corresponding to the inscript and circumscript cylindrical foundations that were obtained by Apsel. The same happens for the vertical and torsional stiffnesses proposed by the other authors. The horizontal and rocking stiffnesses given by Elsabee and Morray do not always

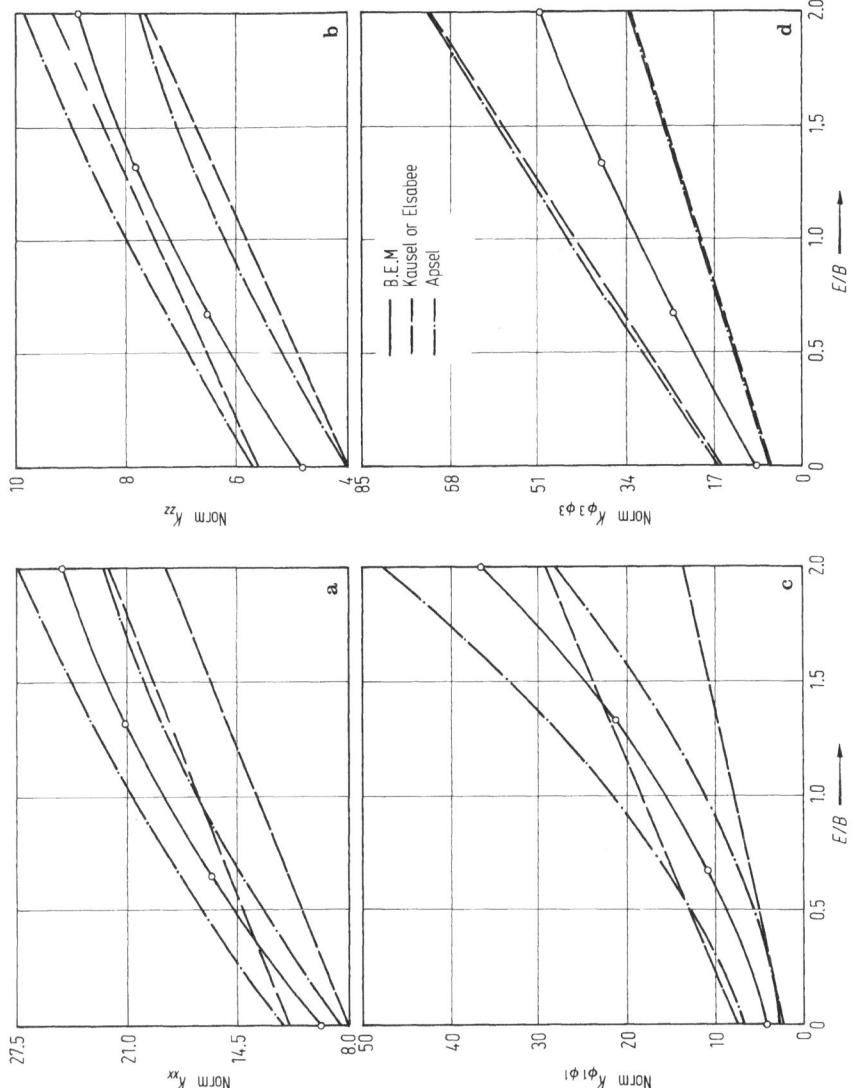

Fig. 12a–d. Static stiffnesses for square embedded foundations

include between them the BEM results, but they are also in desagreement with Apsel's. Figure 12 shows an increase of the stiffnesses with the level of embedment that is slightly lower that the one presented by one the authors in 1978 [17]. This fact may be explained because Dominguez's results were obtained using the single fundamental solution and a soil free-surface discretization extending up to a distance A/B approximately equal to 1 which, as was shown in Figs. 10 and 11, produces a certain over-estimation of the values of the static stiffnesses that increases with the level of embedment.

According to Fig. 12, the following approximate formulae for the static stiffnesses of square embedded foundations may be proposed [18]:

Quadratic approximation:

$$K_{xx} = 9.41 \frac{GB}{2-v} \left\{ 1 + 1.13 \frac{E}{B} - 0.16 \left(\frac{E}{B}\right)^2 \right\},$$

$$K_{zz} = 4.75 \frac{GB}{1-v} \left\{ 1 + 0.47 \frac{E}{B} - 0.05 \left(\frac{E}{B}\right)^2 \right\},$$

$$K_{\phi 1 \phi 1} = 4.38 \frac{GB^3}{1-v} \left\{ 1 + 0.98 \frac{E}{B} + 1.13 \left(\frac{E}{B}\right)^2 \right\},$$

$$K_{\phi 3 \phi 3} = 8.71 GB^3 \left\{ 1 + 2.8 \frac{E}{B} - 0.19 \left(\frac{E}{B}\right)^2 \right\};$$

(54)

Linear approximation:

$$K_{xx} = \frac{47}{5} \frac{RB}{2-v} \left\{ 1 + \frac{9}{10} \frac{E}{B} \right\},$$

$$K_{zz} = \frac{19}{4} \frac{GB}{1-v} \left\{ 1 + \frac{2}{5} \frac{E}{B} \right\},$$

$$K_{\phi 1 \phi 1} = \frac{35}{8} \frac{GB^3}{1-v} \left\{ 1 + \frac{5}{2} \frac{E}{B} \right\},$$

$$K_{\phi 3 \phi 3} = \frac{26}{3} GB^3 \left\{ 1 + \frac{51}{20} \frac{E}{B} \right\},$$

(55)

where the variation with Poisson's modulus has been assumed to be the same of the circular surface foundation.

A study of the static stiffnesses of surface or embedded foundations of any shape may be done using the BEM. In Ref. [17] results were obtained for rectangular surface and embedded foundations.

2.7.4 Dynamic Stiffness Coefficients

Assuming a linear elastic or viscoelastic soil, the values of the dynamic stiffness components may be computed without special difficulties for different values of the dimensionles frequency a_0. In Refs. [17, 18], dynamic stiffness coefficients for square and rectangular, surface and embedded foundations may be found. In the following, only a small number of results taken from those references will be shown.

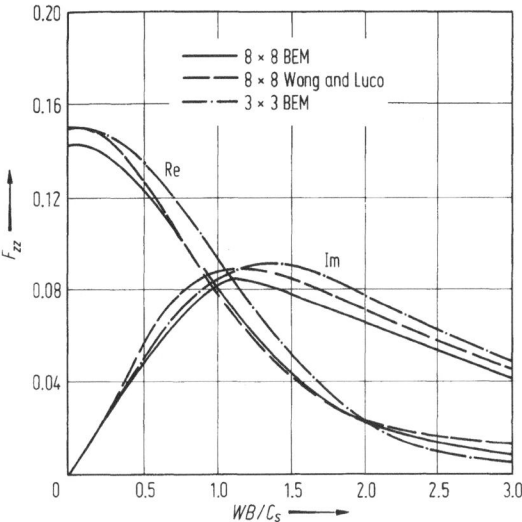

Fig. 13. Vertical compliance for surface square foundations

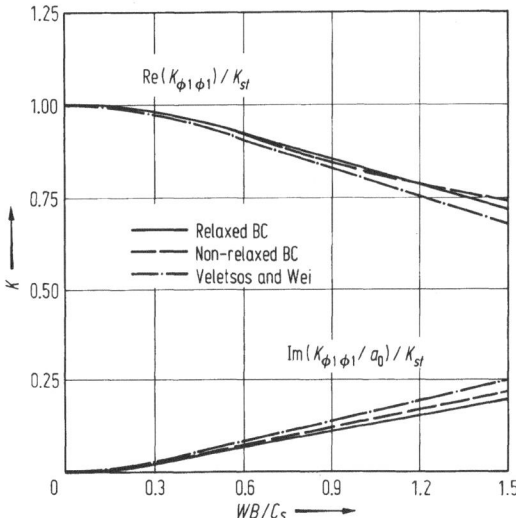

Fig. 14. Rocking stiffness coefficients. Surface square foundations

A comparison of the vertical compliance obtained using two BE meshes and the results published by Wong and Luco [13] is shown in Fig. 13. Rocking stiffness coefficients for a surface square foundation are shown in Fig. 14. Results obtained using relaxed (x and y directions independent of z) and non-relaxed boundary conditions are in good agreement with those computed using the values given by

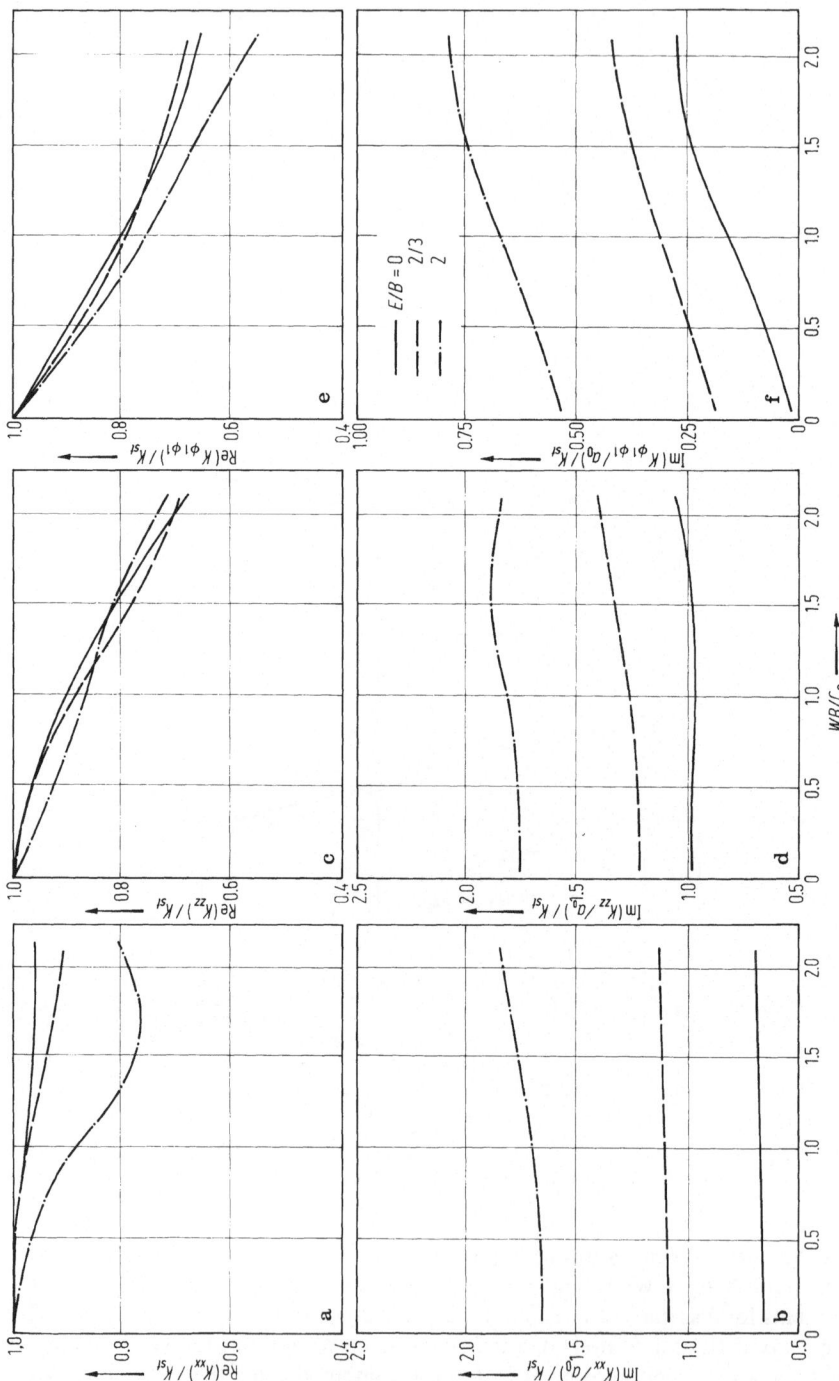

Fig. 15a–f. Dynamic stiffness coefficients. Square embedded foundations

Veletsos and Wei [9] for circular foundations. An equivalent radius of a foundation with the same moment of inertia with respect to the rocking axis was used for comparison.

Figure 15 shows the variation with frequency of the dynamic horizontal, vertical and rocking stiffness coefficients for several values of E/B. Except for the horizontal component, the real parts normalized with respect to the corresponding static values present a variation with a_0 that is almost linear and independent of E/B. The imaginary horizontal and vertical coefficients are highly dependent of E/B but remain almost constant with frequency a_0; i.e. the variation of the imaginary part of the dynamic stiffness is almost linear with a_0.

2.8 Two-Dimensional Foundations

The frequency domain formulation of the BEM for 2-D regions may be used to compute dynamic stiffnesses of strip footings in the same way that has been done for 3-D foundations. The number of unknowns is smaller and the discretization and treatment of the data, simpler. In the following, even though some results for a homogeneous viscoelastic soil model will be analyzed, the main attention will be paid to functions resting on non-homogeneous soils.

Earthquake damage observation shows that local soil mechanical properties, underground and surface topography, and foundation geometry have an important effect on the dynamic behaviour of structures. The complexity of the system to be modeled has made numerical methods the most suitable way to deal with the problem. After the development of energy absorbing boundaries [24, 25], finite elements became a widely used technique for this kind of problems. However, finite element models present two unavoidable requirements that may be important in certain cases. First, the model must be bounded at the bottom by a rigid bedrock and second, the soil away from the foundation must be represented by parallel layers unbounded in the horizontal direction. These two conditions are not always close to reality. There are cases where the base under the soil deposit is not very rigid or the soil geometry is far from being horizontally layered (for example, narrow valleys). The BEM is a good alternative for those problems. It permits an easy representation of soils with irregular shape (Fig. 16) and the modelling of soils bounded at the bottom by a compliant half-space. Some studies of the effect of the existence of a compliant bedrock were done by Gazetas and Roësset [55, 56] using a semi-analytical procedure that is limited to surface foundations on horizontally layered soils.

In the following, compliances of surface massless foundations are presented. The validity of the horizontally layered soil and rigid bedrock finite elements hypothesis are analysed for cases where the soil deposit has elliptical shape and/or the lower half-plane is not infinitely rigid. The compliances of surface strip footings resting on a viscoelastic half-plane are studied first for comparison. Secondly, compliances of foundations resting on the surface of a soil deposit which is on the top of a viscoelastic half-plane model are studied. The analysis is done for a soil layer on top of the viscoelastic half-plane and also for a semielliptical valley included in the

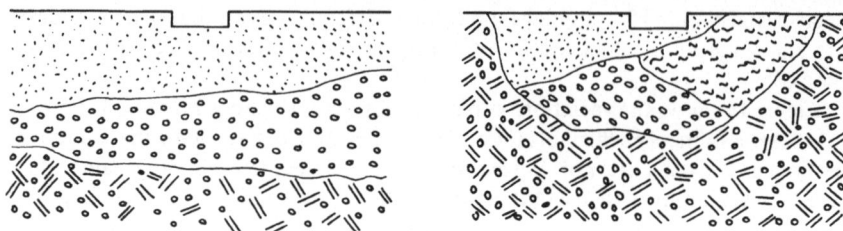

Fig. 16. Zoned soils

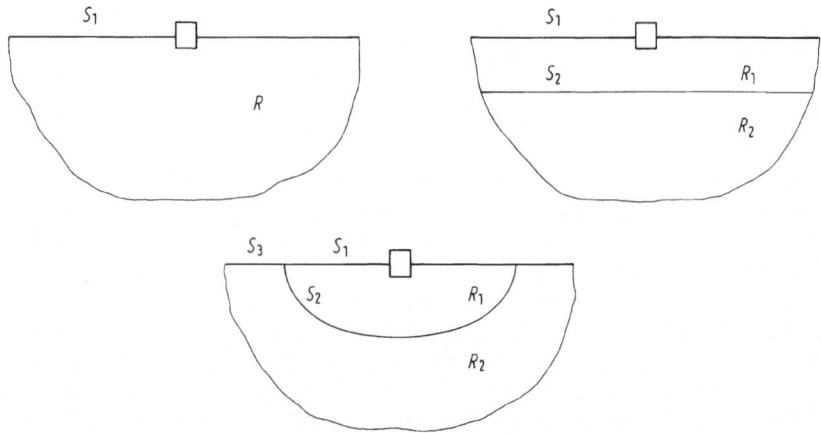

Fig. 17. Homogeneous and zoned viscoelastic soil models

half-plane (Fig. 17). In order to check the importance of a deformable lower bed, the rigidity of the half-plane takes values going from the same of the soil deposit (homogeneous half-plane) to infinity (rigid bedrock). The axis ratio (D/H) of the ellipse takes several values from unity to infinity (horizontal layer) to show up the influence of the soil deposit not being a horizontal layer. A more complete study of the use of the BEM for dynamic stiffnesses of foundations on zoned viscoelastic soils and, of the effects of the shape of the soil deposit and the compliant bedrock is done in Ref. [57].

2.8.1 Viscoelastic Half-plane

To obtain compliances of a surface strip footing resting on a viscoelastic half-plane, half of the soil surface is discretized into 43 elements, ten of which are under the foundation (Fig. 18). An amount of free-field equal to ten times the foundation width is discretized. This amount is not necessary but has been used to maintain the same surface discretization of the layered model. Values of the compliance coefficients are obtained for dimensionless frequencies going from 0 to 2.5. A Poisson's ratio equal to 0.4 and a 5% damping are assumed.

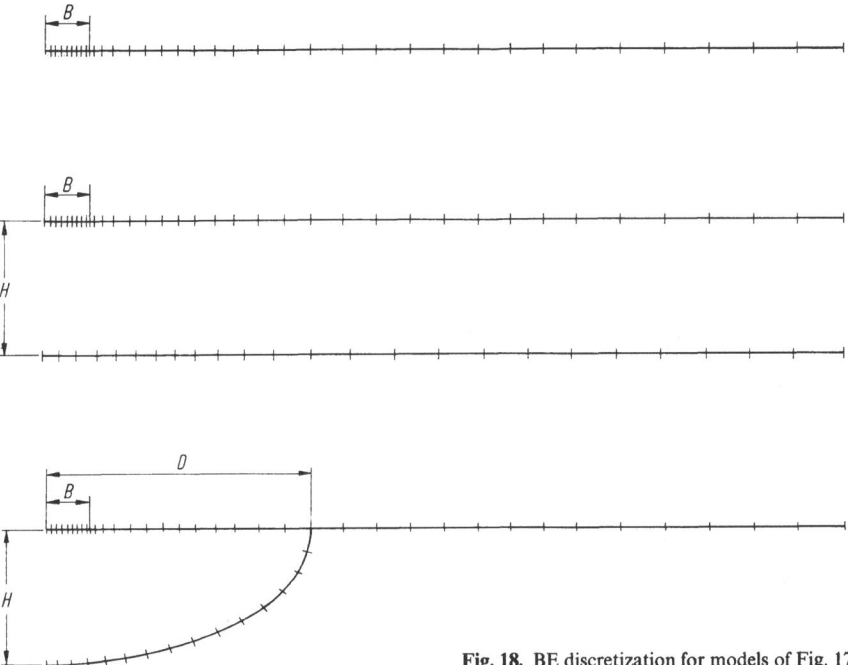

Fig. 18. BE discretization for models of Fig. 17

Figures 19a and b show the computed values of the real and imaginary part of the vertical compliance. The results are compared with those obtained by Gazetas and Roësset [55, 56] by direct integration of the wave equation using a space Fourier transform of a load distributed over small segments under the foundation. A good agreement may be observed in the figures.

2.8.2 Layered Soil. Effect of a Compliant Bedrock

The soil model of Fig. 17b has been used to show the influence of a non-infinitely rigid bedrock on the foundation compliance. A parametric study has been done to show this influence. The ratio of the S wave velocity of the base to the S wave velocity of the soil deposit ($RC_s = C_{s2}/C_{s1}$) was set successively equal to 1 (viscoelastic half-space), 2, 4, and infinity (rigid bedrock).

The real part of the horizontal compliance for a layer depth four times the foundation half-width ($H/B = 4$) and rigid bedrock is presented in Fig. 20a for comparison. The figure shows a good agreement between BEM and Bazetas-Roësset results. A good agreement may also be observed in Fig. 20b for the real part of the rocking compliance when $H/B = 2$ and the bedrock is rigid. It may be seen in Figs. 20a and b how the existence of a soil layer introduces resonance peaks that do not exist for the half-plane.

Figure 20c shows the real part of the horizontal stiffness for $H/B = 4$ and different values of the S wave velocity ratio. It may be seen how the resonance peaks

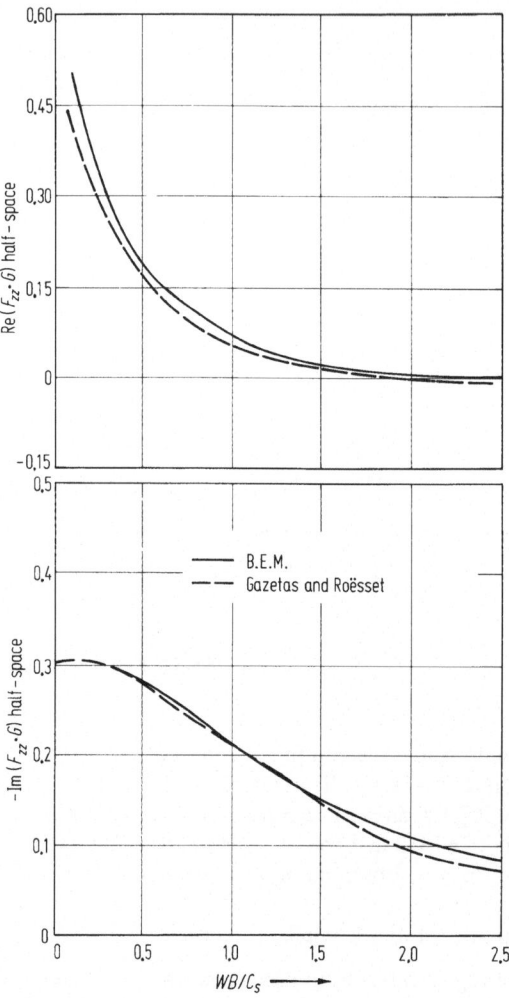

Fig. 19. Vertical compliance

decrease as the bedrock gets more flexible. The peaks appear approximately at natural frequencies of the soil layer given by the monodimensional amplification theory. Thus, the first one is close to the first S waves natural frequency ($a_{s1} = 0.39$); the second peak is in the proximity of the first P waves natural frequency ($a_{p1} = 0.96$); the third is near $a_{s2} = 1.18$; and the forth near $a_{s3} = 1.96$. The imaginary part of the horizontal stiffness, presented in Fig. 20d, also has peaks near the layer natural frequencies. It may be noticed how the low frequency values are close to the percentage of damping.

The real part of the vertical compliance for $H/B = 2$ is shown in Fig. 20e.

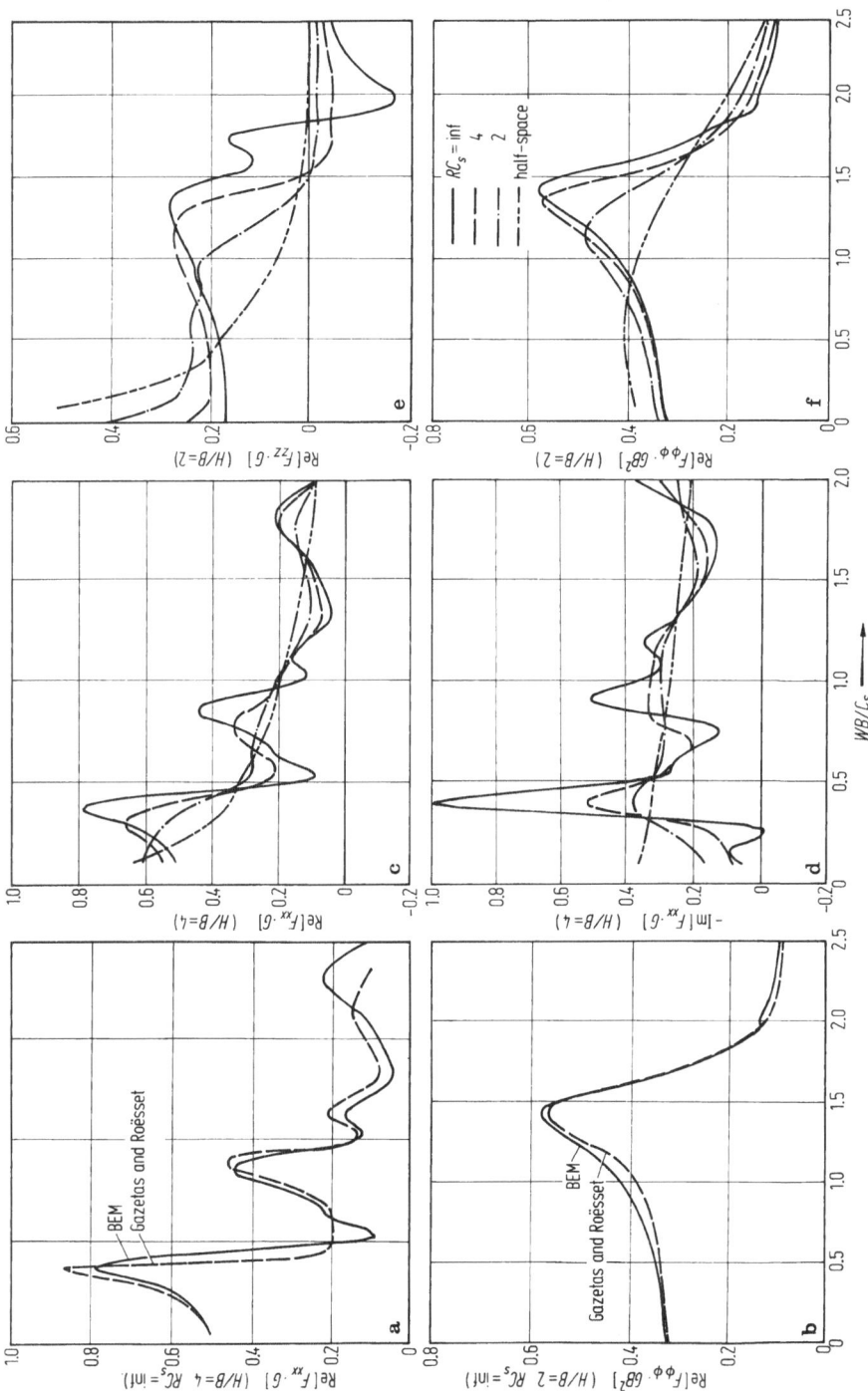

Fig. 20a–f. Compliances for soil layer on a compliant bedrock

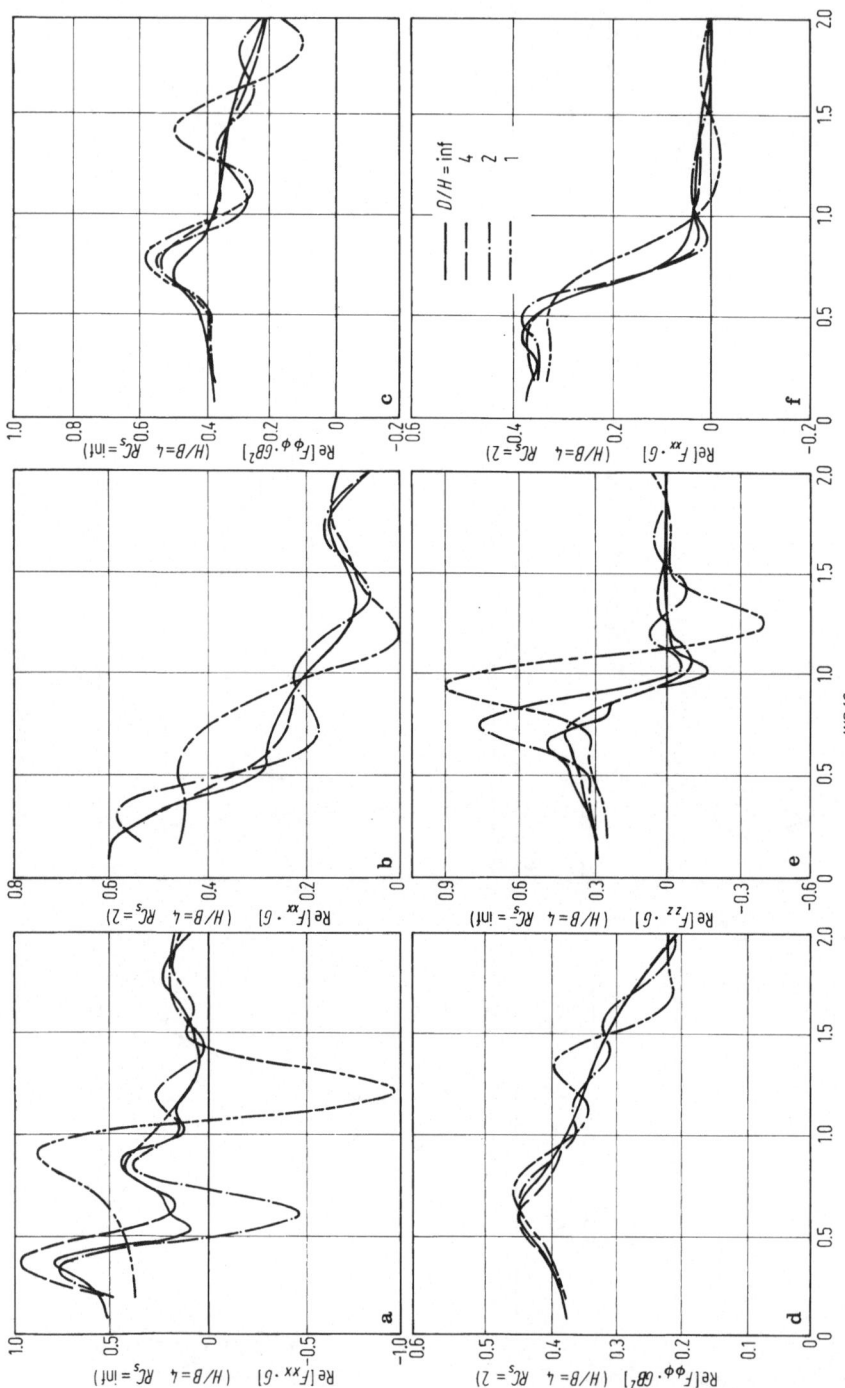

Fig. 21a–f. Effect of the shape of the soil deposit and compliant bedrock on the foundations compliances

2.8.3 Elliptical Soil Deposit on Viscoelastic Bedrock

As we said above, the hypothesis of horizontal soil layers boundless in the horizontal direction may not be in correspondence with reality. However, this hypothesis has to be done in Finite Elements due to the fact that a Consistent Boundary Matrix is obtained using an expansion of the motion in the wave modes of the boundless layers. If the BEM is used, the above limitation does not exist and the arbitrary shape soil profiles may be modeled. In the following an analysis of the influence of the shape of the soil deposit is done using the nodel of Fig. 17c.

The axes of the ellipse are $2D$ and $2H$ respectively. The ratio D/H will take several values from infinity (horizontal boundless layer) to 1 (semicircular soil deposit). The velocity ratio RC_s is considered as another variable. Thus, it will be possible to study the influence of the shape for a given velocity ratio and vice versa.

The real parts of the horizontal, rocking and vertical compliances are shown for several values of D/H in Fig. 21. Figures 21a, c and e were obtained for $RC_s = \infty$, and Figs. 21b, d and f for $RC_s = 2$. It may be noticed in those figures that the horizontal static compliance decreases with D/H and, to a less extent, the same happens to the vertical static compliance. The rocking static compliance does not show noticeable variation with D/H. The figures show an increase in the resonance phenomenon and osillations with frequency, when D/H decreases. This fact may be explained by an increase in the reflexions at the bottom of the soil deposit which is less important when RC_s is small and more energy is transmited to the half-plane.

With respect to the horizontal layer hypothesis, it may be said that for the case presented ($H/B = 4$) a boundless horizontal layer can be assumed without much error for values of D/H greater than 4 while for smaller values that hypothesis may lead to important errors. In all cases the rocking coefficient is the less influenced by D/H being the horizontal compliance the one that varies more with that parameter.

It has been shown in this section that the hypothesis of layered soil and rigid bedrock may lead to erroneous values of the compliances if the base is not very rigid or the soil deposit is not wide enough.

The problems analysed are examples of the capabilities of the BEM for the computation of dynamic stiffnesses of foundations.

2.9 Axisymmetric Foundations

The first formulation of the BEM for axisymmetric elastostatic problems was done in 1975 [58, 59]. The integral representation in cylindrical coordinates has the same expression of 2-D and 3-D problems when one uses a fundamental solution corresponding to ring loads following the radial, tangential and axial direction, respectively. Those fundamental solutions are written in terms of Legendre functions or elliptic integrals [59, 60, 46], which makes their integration along the boundary elements rather involved. The harmonic ring load fundamental solution may be obtained in terms of an infinite line integral of Hankel functions [46] and its integration along the elements is again complicated. On the contrary, the 3-D static or harmonic point load solution may be easily integrated over axisymmetric surface elements, and they are used here for the BEM treatment of axisymmetric problems.

The basic BEM equation for steady-state harmonic 3-D problems, is written in cylindrical coordinates using matrix notation

$$C^c(\xi)u^c(\xi) = \int_\Gamma Q^T(\xi)\,U(x,\xi)Q(x)t^c(x)\,d\Gamma(x)$$

$$- \int_\Gamma Q^T(\xi)\,T(x,\xi)Q(x)u^c(x)\,d\Gamma(x), \qquad (56)$$

where the superscript "c" stands for cylindrical.

$$u(x) = Q(x)u^c(x),$$
$$u(\xi) = Q(\xi)u^c(\xi), \qquad (57)$$

the same being applicable to $t(x)$, $t(\xi)$

$$C^c(\xi) = Q^T(\xi)C(\xi)Q(\xi) \qquad (58)$$

and

$$Q(x) = \begin{pmatrix} \cos\theta(x) & -\sin\theta(x) & 0 \\ \sin\theta & \cos\theta & 0 \\ 0 & 0 & 1 \end{pmatrix}. \qquad (59)$$

The transformation of tensors U and T are in agreement with a more genèral study of the transformation of these tensors done by Rizzo and Shippy [61].

Since ξ is the collocation point selected to apply the unit point load, it may be assumed that $\theta(\xi) = 0$ (Fig. 22), which makes the transformation matrix $Q(\xi) = I$. The kernels of the integrals have the form

$$U^c = U(x,\xi)Q(x) = \begin{pmatrix} U_{11}\cos\theta + U_{21}\sin\theta & -U_{11}\sin\theta + U_{12}\cos\theta & U_{13} \\ U_{21}\cos\theta + U_{22}\sin\theta & -U_{21}\sin\theta + U_{22}\cos\theta & U_{23} \\ U_{31}\cos\theta + U_{32}\sin\theta & -U_{31}\sin\theta + U_{22}\cos\theta & U_{33} \end{pmatrix}, \qquad (60)$$

where $\theta = \theta(x)$.

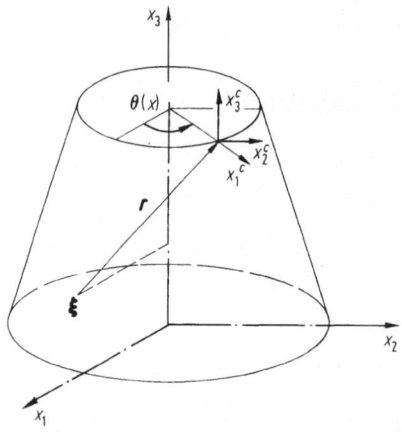

Fig. 22. Cylindrical coordinates

Equation (56) may be written in short form as

$$C^c u^c(\xi) = \int_\Gamma U^c t^c(x)\, d\Gamma - \int_\Gamma T^c u^c(x)\, d\Gamma. \tag{61}$$

One half of the meridional section of the body is discretized into elements. Displacements and tractions are independent of θ and assumed to be constant (linear, quadratic, etcetera) along each line element. Thus, Eq. (61) may be written for boundary node p as

$$C_p u_p^c + \left\{ \sum_{q=1}^Q \int_{\Gamma_q} \rho \psi\, d\Gamma_q \int_0^{2\pi} T^c\, d\theta \right\} u_q^c$$

$$= \left\{ \sum_{q=1}^Q \int_{\Gamma_q} \rho \psi\, d\Gamma_q \int_0^{2\pi} U^c\, d\theta \right\} t_q^c, \tag{62}$$

where Γ_q are boundary elements of the section and ψ shape functions along line elements. The integration along θ may be easily done by numerical procedures.

$$C_p u_q^c + \sum_{q=1}^Q H_{pq} u_q^c = \sum_{q=1}^Q G_{pq} t_q^c. \tag{63}$$

The matrices H_{pq} and G_{pq} that relate two nodes p and q have the pattern:

$$\begin{pmatrix} * & 0 & * \\ 0 & * & 0 \\ * & 0 & * \end{pmatrix} \begin{matrix} \leftarrow \rho \\ \leftarrow \theta, \\ \leftarrow z \end{matrix} \tag{64}$$

where the zeros denote elements that are null due to the skewsymmetry of the corresponding terms in Eq. (60). It is clear that the torsion and the radial-axial problems are uncoupled and both may be studied over a 2-D domain.

When the boundary conditions are not axisymmetric (for instance, horizontal or rocking motion of cylindrical foundations), the problem may still be analysed by means of a plane model. The problem is divided into a number of plane uncoupled situations by representing the prescribed loading or displacements by a Fourier series along the tangential coordinate [62]. Each term of the series produces displacements and stresses in the same Fourier mode and if the prescribed values do not vary very rapidly around the axis, a few modes will be enough for an accurate solution. The Fourier expansion is of the form

$$u_\rho = \sum_{n=0}^\infty (u_{n\rho}^s \cos n\theta + u_{n\rho}^a \sin n\theta),$$

$$u_\theta = \sum_{n=0}^\infty (-u_{n\theta}^s \sin n\theta + u_{n\theta}^a \cos n\theta), \tag{65}$$

$$u = \sum_{n=0}^\infty (u_{nz}^s \cos n\theta + u_{nz}^a \sin n\theta),$$

where "s" indicates the symmetric terms and "a" the antisymmetric ones.

For each Fourier mode amplitude a discretized boundary equation like Eq. (63) may be written with the only difference that H_{pq} and G_{pq} are now obtained by integration of T^c and U^c weighted by a sine or cosine function.

$$\int_0^{2\pi} U^c \sin n\theta \, d\theta; \quad \int_0^{2\pi} U^c \cos n\theta \, d\theta. \tag{66}$$

It is worth noting that since $\sin n\theta$ has a zero value for $\theta = 0$, the ξ point cannot be located at $\theta(\xi) = 0$ to compute the amplitude of those terms of the series. One only has to move ξ to a point; for instance, $\theta(\xi) = -\pi/2n$, where the amplitude is not zero. This change is easily taken into account by a shift of the origin of θ in Eq. (60).

Vertical and torsional displacements of cylindrical foundations imply axisymmetric boundary conditions and the corresponding stiffnesses are computed using the basic equation (Eq. 62). Horizontal and rocking displacements of the foundation imply displacement fields of the form

$$u_\rho = u_{1\rho} \cos \theta,$$

$$u_\theta = -u_{1\theta} \sin \theta, \tag{67}$$

$$u_z = u_{1z} \cos \theta,$$

which makes necessary to take $\theta(\xi) = -\frac{\pi}{2}$ for the second component equations.

The integrals may be easily done numerically except when working in the same annular area containing the collocation point. In those cases, the static fundamental solution is subtracted from the harmonic point load solution. The difference does not contain any singularity and can be integrated numerically. The static solution is integrated analytically along the meridional line and numerically along the two extreme circunferences of each element.

A special numerical integration scheme is successfully used for axisymmetric problems [20]. When the fundamental solution is integrated around the axis, a Gauss quadrature may be applied to every semi-ring; however, its accuracy is easily improved by increasing the density of integration points near the collocation point by means of a parabolic transformation of the circunferencial coordinate

$$\theta = \frac{\pi}{4}(\eta + 1)^2; \quad -1 \leqslant \eta \leqslant 1. \tag{68}$$

The above formulation is used to compute dynamic stiffnesses of circular foundations. Figure 23 shows the variation of the horizontal stiffness coefficients with frequency for a circular foundation on an elastic half-space ($\nu = 1/3$). Only 8 constant elements are used. Results are in good agreement with those given by Veletsos and Wei [9].

A circular foundation on a layered half-space has also been studied using the above BEM formulation [9, 63]. The soil properties are the same used by Chapel [64], given in Fig. 24.

Values of the vertical stiffness coefficients are compared in Fig. 25 with Chapel's and Luco's [65] results. The BE model used consist of 8 constant elements under the footing, 7 constant elements on the soil free-surface and 8 elements along the internal boundary. As may be seen in the figure, the agreement is particularly good with Luco's results.

The direct formulation of the BEM for axisymmetric problems may be applied to surface or embedded foundations and homogeneous or zoned soils with irregular

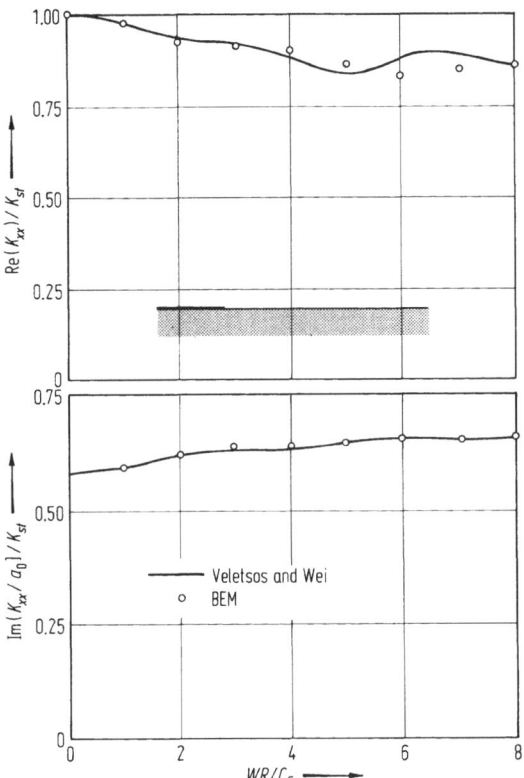

Fig. 23. Horizontal stiffness coefficients for a rigid disk. (eight elements under the footing. No elements on the soil free-surface)

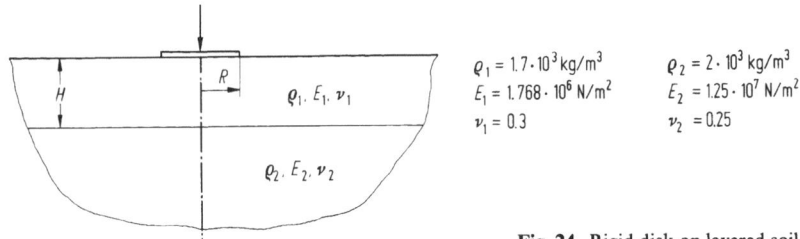

$\varrho_1 = 1.7 \cdot 10^3 \, \text{kg/m}^3$ $\varrho_2 = 2 \cdot 10^3 \, \text{kg/m}^3$

$E_1 = 1.768 \cdot 10^6 \, \text{N/m}^2$ $E_2 = 1.25 \cdot 10^7 \, \text{N/m}^2$

$v_1 = 0.3$ $v_2 = 0.25$

Fig. 24. Rigid disk on layered soil

subsurface topography. An indirect formulation for horizontally layered soils was used by Apsel [21] to obtain dynamic stiffnesses. The use of a fundamental solution for the layered soil permits a discretization only of the soil foundation interface but the computation and integration of the solution is more involved than the simple 3-D complete space solution.

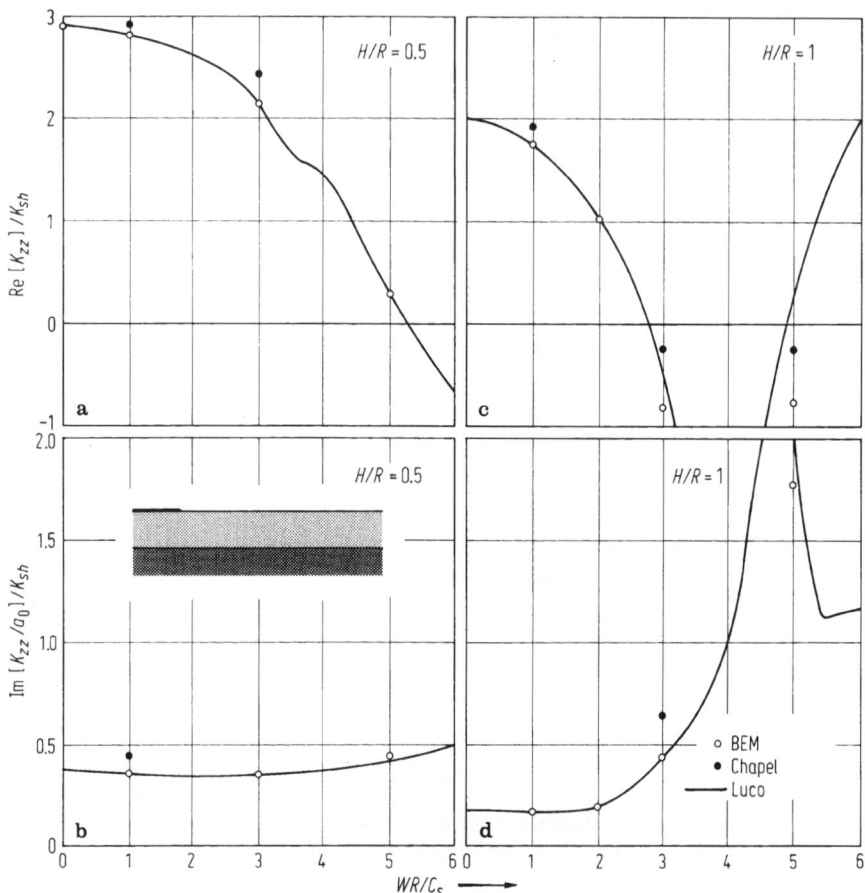

Fig. 25a–d. Vertical stiffness of a disk on a layered half-space

2.10 Seismic Response of Foundations

Diffraction problems dealing with infinite or semiinfinite regions are usually formulated decomposing the total displacement and stress fields into to parts. One is
the undisturbed (free) field $(u^{(f)}, \sigma^{(f)})$ and the other the scattered field $(u^{(s)}, \sigma^{(s)})$. This
decomposition permits the use of the displacements integral representation of the
scattered field for an external region, which consist of integrals that only extend
over the internal boundaries since the radiation and regularity conditions [41] are
satisfied.

The integral representation of the scattered field for points on the soil free-
surface or on the soil-foundation interface when the soil is a homogeneous viscoelastic half-space may be written, following Eq. (46), as

$$\tfrac{1}{2}[\,u_k(\xi) - u_k^{(f)}(\xi)] = \int_{S_1} U_{ik}[t_i - t_i^{(f)}]\,dS - \int_{S_1} T_{ik}[u_i - u_i^{(f)}]\,dS, \tag{69}$$

where S_1 represents (Fig. 17) the soil-foundation interface plus the soil free surface, $u_i^{(f)}$ and $t_i^{(f)}$ are known and all the other symbols have the usual meaning. Once S_1 has been discretized and the boundary conditions applied, the resulting system of equations gives the unknown values u_i and t_i.

For non-homogeneous half-spaces the $u_i^{(f)}$, $t_i^{(f)}$ exists only for the outermost zone and corresponds to a uniform half-space with the properties of that zone. For instance, in Figs. 17b and c the total fields are:

$$
\left.
\begin{aligned}
u(x) &= u^{(f)}(x) + u^{(s)}(x) \\
\sigma(x) &= \sigma^{(f)}(x) + \sigma^{(s)}(x)
\end{aligned}
\right\} x \in R_2,
$$

$$
\left.
\begin{aligned}
u(x) &= u^{(s)}(x) \\
\sigma(x) &= \sigma^{(s)}(x)
\end{aligned}
\right\} x \in R_1.
\tag{70}
$$

The system of equations for the soil models of Figs. 17b and c, once all the boundaries are discretized, is composed of: the integral representation of the scattered displacements for nodes on S_2 (Fig. 17b), or on $S_2 + S_3$ (Fig. 17c), as boundaries of zone R_2, plus the integral representation of the total displacements for nodes on $S_1 + S_2$ as boundaries of zone R_1. The scattered fields for points on S_2 or S_3 as part of R_2, are written as the difference between the total fields and the known free-fields. Compatibility and equilibrium conditions of the total fields are established along the internal boundary (S_2) and boundary conditions are prescribed for the total displacements or tractions along the external boundaries (S_1 or $S_1 + S_3$). Thus, the total displacements and tractions over the boundaries can be computed by means of the system of equations.

The motion of a rigid massless foundation induced by incident waves is computed following two steps. For the first step, zero tractions at the free-surface and zero displacements under the footing are prescribed. The solution of the system gives the tractions under the footing and its resultant R may be easily computed. The second step is the determination of the rigid body motion of the footing by solving the system $Ku^r = -R$, where K is the foundation stiffness matrix that is required for the soil-structure interaction analysis and in any case may be computed with little extra effort using the same integrals along the elements (H_{pq} and G_{pq}) of the first step.

The free field motions are derived from potentials [18] or directly defined in terms of exponential functions [37]. The tractions are obtained by differentiation of those displacements.

2.10.1 Three-Dimensional Foundations

Figure 26 shows the amplitude of the horizontal motion due to SH waves impinging on a square foundation with an angle $\phi = 0, 45$ and 90 degrees with the x-axis. The foundation rests on a linear elastic half-space with a Poisson's modulus $v = 1/3$. Results are compared with those reported by Wong and Luco [34] for the same angle. A good agreement between both solutions is observed. The constant elements model used is also shown in the figure.

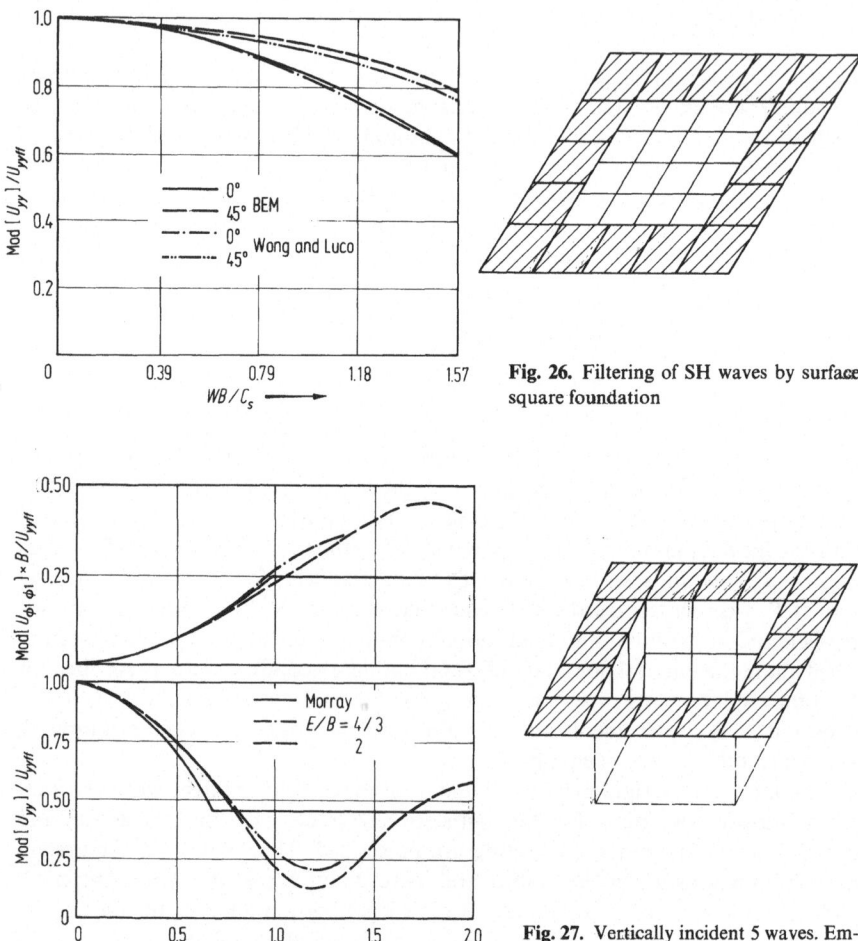

Fig. 26. Filtering of SH waves by surface square foundation

Fig. 27. Vertically incident S waves. Embedded foundations

The response of square embedded foundations to vertically propagating S waves is presented in Fig. 27. The results are plotted versus the natural frequency of the layer of soil corresponding to the embedment, $f_0 = C_s/4E$. The values of the horizontal and rocking motions are compared with the approximate formulae proposed by Morray [53] for circular foundations embedded in a soil stratum. An equivalent radius $R_e = \sqrt[4]{16/3\pi}\, B$ was taken. The comparison of both kinds of results is good, and the oscillations of the values around the approximate solution were also in Morray's results.

A parametric study of the response of embedded square foundations to travelling SH, SV and P waves may be found in the work by Dominguez [37].

2.10.2 Two-Dimensional Foundations

The three soil models of Fig. 17 are used again. The boundary discretization are also the same of Fig. 18. The lateral walls of embedded foundations are discretized using elements of the same size of those on the bottom. The soil properties are the same used as in Sect. 2.8.

Viscoelastic Half-plane

The amplitude of the vertical displacement of a strip footing for P waves and two different angles of incidence ($\theta = 0°$ and $60°$) are shown in Fig. 28. The angle is measured with respect to the vertical direction. Displacements are referred to the undisturbed free-field at $x = z = 0$. An important reduction of the displacement with E/B may be seen in Fig. 28.

For non-vertically incident waves ($\theta = 60°$) the effect of the embedment is smaller than for vertical waves. However, the surface foundation displacement amplitude is now frequency dependent. A minimun may be seen for $E/B = 2$. According to the one-dimensional theory, a minimun could be expected for a length of the wave in the vertical direction close to $\lambda = 2E$ which corresponds to $a_0 = 1.81$. This value is close to the minimun of Fig. 28. The minimun for $E/B = 4/3$ would be around $a_0 = 2.71$. Results for the three components of the foundation displacement, and SV and P waves with several different angles of incidence may be found in Ref. [66].

Layered Soil

The soil model of Fig. 17b and the boundary discretization of Fig. 18b are used again. A layer depth $H/B = 4$ is assumed.

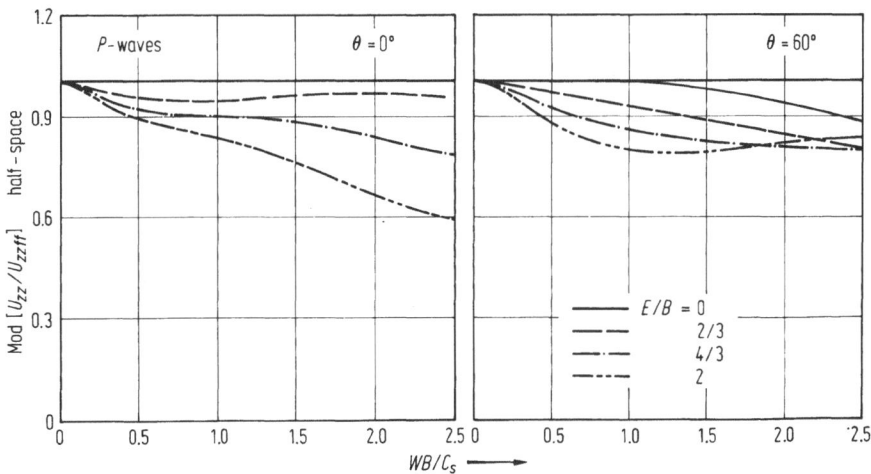

Fig. 28. Vertical displacements for incident P-waves

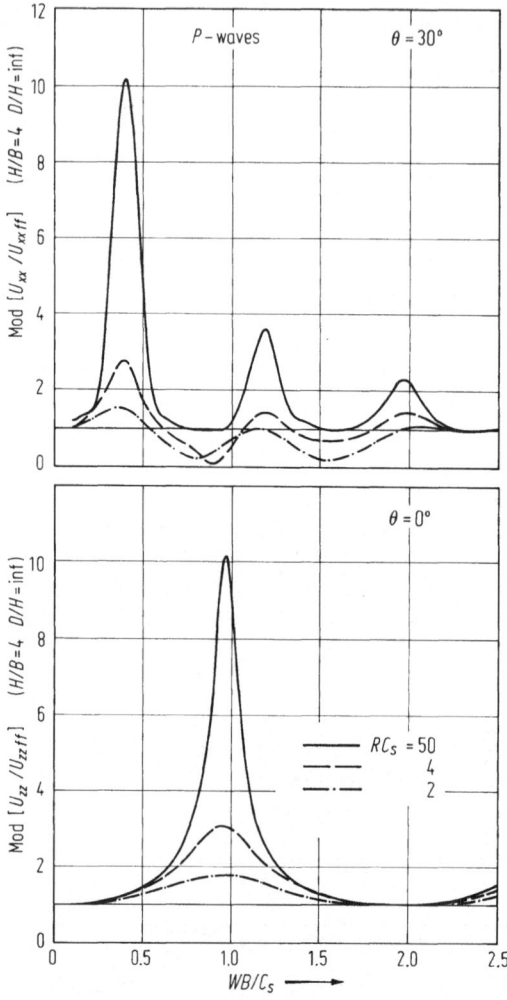

Fig. 29. Displacements of strip footing on layered soil

Figure 29 shows the rigid body horizontal and vertical displacements of a surface strip foundation for incident P waves. Motions are referred to the free field motion at the surface of a half-plane, with the properties of R_2, under the action of waves of the same amplitude. The angle of incidence θ is measured at region R_2 with respect to the normal to the free-surface. Results are presented for three different stiffness ratios: $RC_s = 2$, $RC_s = 4$ and $RC_s = 50$, being the latter an approximation of the rigid bedrock conditions. Figure 29a shows, for $RC_s = 50$, resonance peaks of the horizontal displacement that are located at the S waves natural frequencies of the layer predicted by the one-dimensional theory ($a_{s1} = 0.39$, $a_{s2} = 1.18$ and $a_{s3} = 1.96$). Similarly, a peak appears for the vertical displacement at the first P natural frequency of the layer ($a_{p1} = 0.96$). Figure 29b shows this peak for an angle of

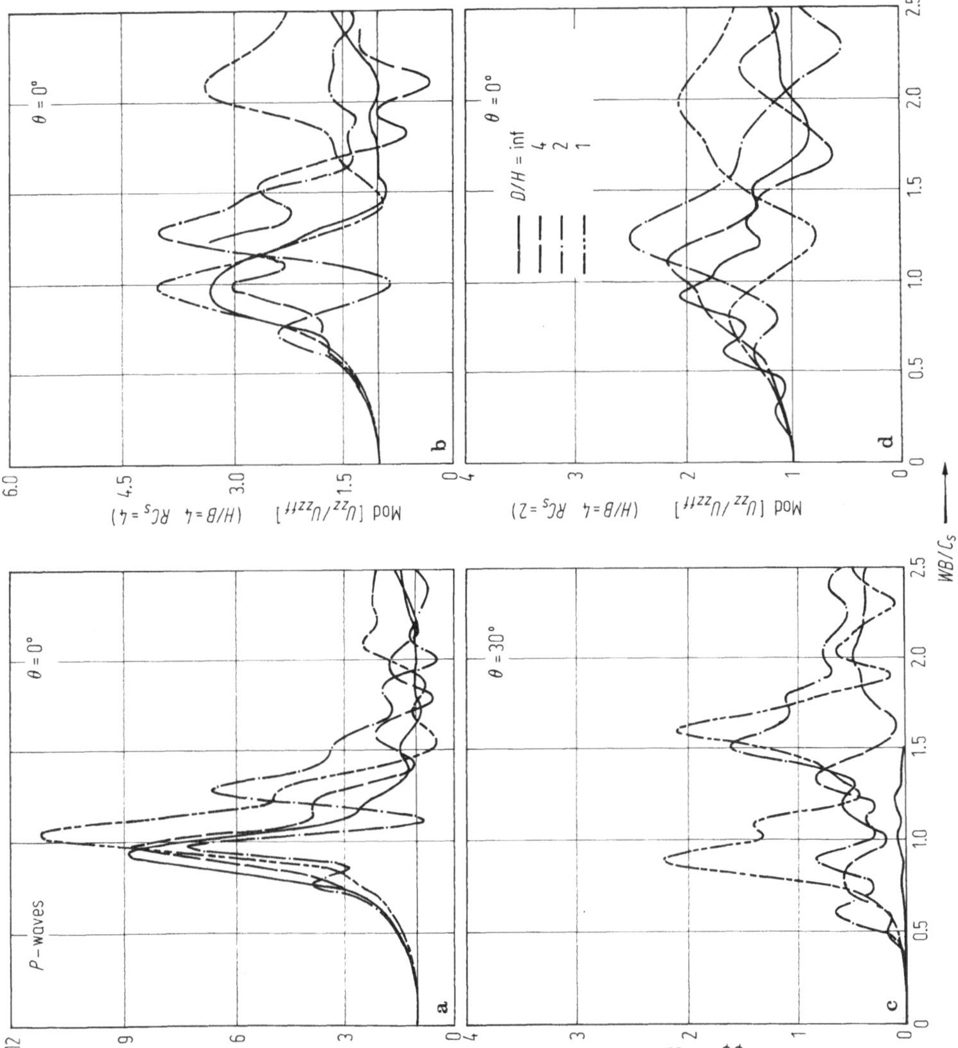

Fig. 30a–d. Displcements of strip footing on elliptical soil deposit

incidence $\theta = 0°$. A reduction of the u_x and u_z amplitudes when the stiffness of the base decreases may be seen in Fig. 29.

The above results show that the relative stiffness between the soil layer and the half-plane has a great influence on the motion of foundations resting on the soil surface.

Elliptical Soil Deposit

The effect of the width of the soil deposit under the footing may be analysed in the same way that it was done for the foundation stiffnesses. Values of the vertical and rocking displacements for rigid ($RC_s = C_{s2}/C_{s1} = 50$) and compliant ($RC_s = 4$ and 2) bedrock are presented in Fig. 30. A reduction of the displacements amplitudes with RC_s may be observed. On the other hand, an increase of the amplitudes when the soil deposit narrows is apparent for most frequencies.

It should be said that there are important variations on the foundation behaviour when the soil profile changes or the relative stiffness RC_s takes different values. Since those variations are not easily predicted, the analysis should be done with a model representing as realistically as possible the soil geometry and properties. The BEM has been shown to be very adequate to generate such models for many soil properties, and foundation and soil geometries.

A more complete parametric study of the foundation displacements due to waves travelling through non-homogeneous soils may be found in Refs. [18, 38].

2.11 Dynamics of Foundations by the Time Domain BEM

Response of foundations to external exitations or travelling waves may be computed using the time domain formulation presented in Sect. 2.4.

The compatibility of displacements of the soil-structure interface elements, when dynamic stiffnesses of foundations are computed in the frequency domain, may be easily prescribed through the boundary conditions. The resultant of the tractions are also computed once the BEM system of algebraic equations has been solved. However, in order to compute the time response for a certain load, it is very convenient to introduce the compatibility and equilibrium conditions in the system of equations

$$u^r = Lu, \tag{71}$$

$$R = Mt, \tag{72}$$

where u^r consist of the components of the motion of the foundation, R is the resultant of forces over the foundation, L, and M are transformation matrices and, u and t consist of the displacements and tractions vectors, respectively, of the nodes on the soil-foundation interface. Combination of Eqs. (71), (72) and (40) result in a system of linear algebraic equations that solve the problem of computing u^r for a given R. The transformations of Eqs. (71) and (72) can obviously be also used in the frequency domain formulation.

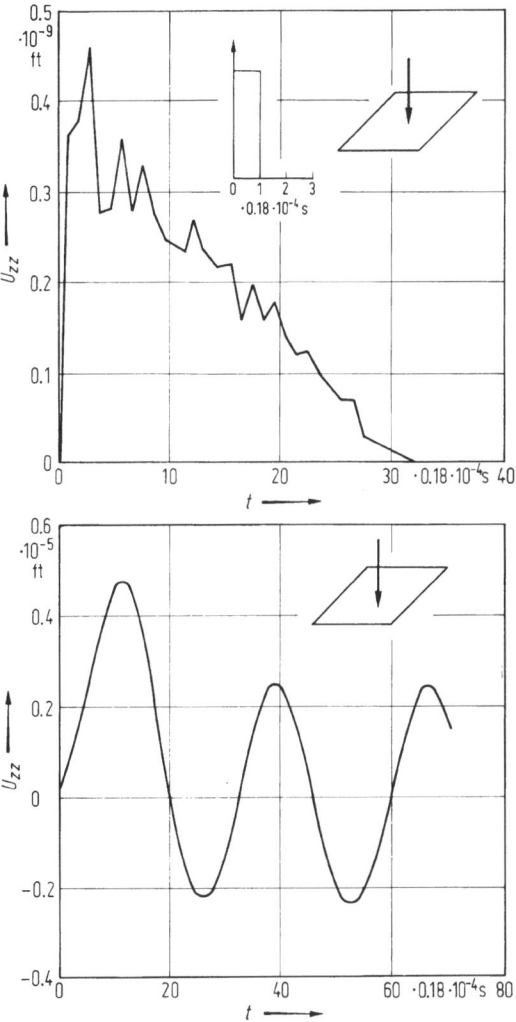

Fig. 31. Time domain response of rigid foundation

In Fig. 31 some of the results presented by Karabalis and Beskos [22] for surface square foundations are reproduced. The vertical displacement of a 5′ × 5′ massless rigid foundation is represented versus time. The properties of the elastic half-space under the footing are: $E = 2.58984 \cdot 10^9$ lb/ft^2, $v = 1/3$ and $\rho = 10.368$ lb·s^2/ft^4. The first figure shows the response to a vertical impulse of 100 Kips and the second, the response to a harmonic vertical load $P = 180 \sin(13000\,t)$ Kips. The Δt used for Eq. (40) was such that the influence of an impulse at a certain node along a time increment n only extend, during the same time increment, over the element to which the node belongs.

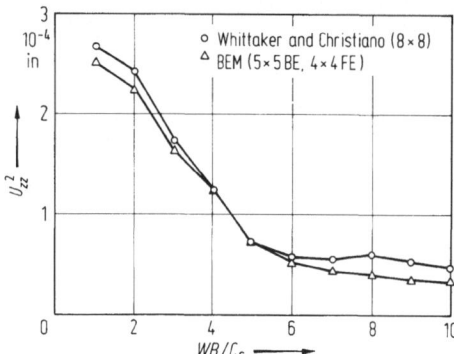

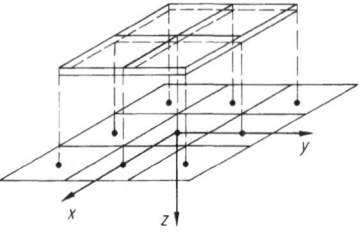

Fig. 32. Vertical displacement at mid-point. Flexible foundation

For flexible foundations, the plate may be discretized into Finite Elements and coupled with the corresponding BE of the soil-foundation interface. In this case, the compatibility and equilibrium conditions [Eqs. (71) and (72)] are substituted by

$$F^e = M't + K_p u, \qquad (73)$$

where K_p is the FE stiffness matrix of the plate where the rotations degrees of freedom have been condensed, F^e consist of the external forces applied to each node and M' transforms tractions into element forces.

Figure 32 shows the results presented by Karabalis, Spyrakos and Beskos [23] for a square flexible massless foundation under the action of a vertical harmonic force applied to the mid-point of the plate. The side of the plate is 5 feet long and the soil properties are the same of the previous example. Even though, the load is harmonic, the problem was solved using the time domain BEM. The results shown in Fig. 32 are in good agreement with Whittaker and Christiano [67].

References

1 Whitman, R.V. and Richart, F.E. "Design procedures for dynamically loaded foundations", Jour. Soil Mech. Fnd. Engrg. ASCE, 93, SM6, 1967, p. 169.
2 Richart, F.E., Woods, R.D. and Hall, J.R. "Vibrations of soils and foundations", Prentice-Hall, 1970.
3 Reissner, E. "Stationäre, axialsymmetrische, durch eine schüttelnde Masse erregte Schwingung eines homonenen elastischen Halbraumes". Ingenieur Archiv., Vol. 7, 1936, pp. 381–396.
4 Arnold, R.N., Bycroft, G.N. and Warburton, G.B. "Forced Vibration of a Body on a Infinite Elastic Solid". Jour. Appl. Mech., ASME, Vol. 22, No. 3, 1965, pp. 391–400.
5 Bycroft, O.N. "Forced Vibration of a Rigid Circular Plate on a Semi-infinite Elastic Space or a Elastic Stratum". Phil. Trans. Royal Soc. of London, 248, 1956, pp. 327–368.
6 Kobori, T., Minari, R. and Suzuki, T. "Dynamic ground compliance of rectangular foundation on an elastic stratum", Proc. 2nd Japan Nat. Symp. on Earthquake Engrg., 1966, pp. 261–266.
7 Collins, W.D. "The Forced Torsional Oscillations of an Elastic Half-Space and Elastic Stratum". Proc. London Math. Soc., 12, No. 46, 1962, pp. 226–244.
8 Paul, H.S. "Vibration of a Rigid Circular Disk on an Infinite Elastic Plate". Jour. Acoust. Soc. Am., 42, 1967, pp. 412–416.
9 Veletsos, A.S. and Wei, Y.T. "Lateral and Rocking Vibration of footings". Jour. Soil Mech. Found. Div., ASCE, 97, 1971, pp. 1227–1248.

10 Luco, J.E. and Westmann, R.A. "Dynamic Response of Circular Foundations". Jour. Eng. Mech. Div., ASCE, 97, 1971, pp. 1381-1395.

11 Veletsos, A.S. and Verbic, B. "Vibration of Viscoelastic Foundation". Jour. Geo. Eng. Div., ASCE, 100, 1973, pp. 225-246.

12 Kobori, T., Minai, R. and Suzuki, T. "The dynamic ground compliance of a rectangular foundation on a viscoelastic stratum". Bull. Disaster Prev. Res. Inst., Kyoto Univ., Japan. 1971.

13 Wong, H.L. and Luco, J.E. "Dynamic Response of Rigid Foundations of Arbitrary Shape", Earthquake Engineering and Structural Dynamics, 4, 1976.

14 Lamb, H. "On the Propagation of Tremors over the Surface of a Elastic Solid". Philosophical Trans. Royal Society of London, Series A, 203, 1904.

15 Elorduy, J., Nieto, J.A. and Szekely, E.M. "Dynamic Response of Bases of Arbitrary Shape Subjected to Periodic Vertical Loading". Proc. International Symp. on Wave Propagation and Dynamic Properties of Earth Materials, 1967.

16 Kitamura, Y. and Sakurai, S. "Dynamic Stiffness for Rectangular Rigid Foundations on a Semi-infinite Elastic Medium", International Jour. of Num. and Analytical Methods in Geomechanics, 1978.

17 Domínguez, J. "Dynamic Stiffness of Rectangular Foundations". Research Report R78-20, Department of Civil Engineering, Massachusetts Institute of Technology, Cambridge, Mass., Aug., 1978.

18 Abascal, R., "Estudio de Problemas Dinámicos en Interacción Suelo-Estructura Mediante el Método de los Elementos de Contorno", Thesis to the Escuela Superior de Ingenieros Industriales, at Seville, Spain, Fev., 1984.

19 Abascal, R. and Domínguez, J. "Dynamic Behavior of Strip Footings on Non-homogeneous Viscoelastic Soils". Int. Symposium on Dynamic Soil-Struct. Int., Minneapolis. Balkema. Roterdam, 1984.

20 Gomez-Lera, S., Domínguez, J. and Alarcón, E. "On the Use of 3-D Fundamental Solutions for Axisymmetric Steady-State Dynamic Problems". Proc. of the 7th Int. Conference on BEM in Engineering. Edt. C.A. Brebbia, 1985.

21 Apsel, R.J. "Dynamic Green's Functions for Layered Media and Applications to Boundary-Value Problems", Thesis presented to the University of California, at San Diego, 1979.

22 Karabalis, D.L. and Beskos, D.E. "Dynamic Response of 3-D Rigid Surface Foundations by Time Domain Boundary Element Method". Earthquake Engineering and Structural Dynamics, Vol. 12, No. 1, 1984, pp. 73-94.

23 Karabalis, D.L., Spyrakos, C.C. and Beskos, D.E. "Dynamic Response of Surface Foundations by Time Domain BEM". Int. Symposium on Dynamic Soil-Struct. Int., Minneapolis. Balkema, Roterdam, 1984.

24 Waas, G. "Linear Two-Dimensional Analysis of Soil Dynamics Problems in Semi-Infinite Layered Media". Thesis presented to the University of California, at Berkeley, 1972.

25 Kausel, E. "Forced Vibrations of Circular Foundations on Layered Media". Research Report R74-11, Civil Engineering Department, Massachusetts Institute of Technology, Cambridge, Mass., 1974.

26 Trifunac, M.D. "Surface Motion of a Semicylindrical Alluvial Valley for Incident Plane SH Waves", Bulletin of the Seismological Society of America, Vol. 61, 1971, pp. 1755-1770.

27 Trifunac, M.D. "Scattering of Plane SH Waves by a Semicylindrical Canyon", Earthquake Eng. and Stru. Dynamics, Vol. 1, 1973, pp. 267-281.

28 Wong, H.L. and Trifunac, M.D. "Scattering of Plane SH Waves by a Semi-Elliptical Canyon", Earthquake Engineering and Structural Dynamics, Vol. 3, 1974, pp. 157-169.

29 Wong, H.L. and Jennings, P.C. "Effect of Canyon Topography on Strong Ground Motion", Bulletin of the Seismological Society of America, Vol. 65, 1975, pp. 1239-1257.

30 Sánchez-Sesma, F.J. and Rosenblueth, E. "Ground Motion at Caynons of Arbitrary Shapes Under Incident SH Waves", Earthquake Engineering and Structural Dynamics, Vol. 7, No. 5, 1979, pp. 441-450.

31 Wong, H.L. "Diffraction of P, SV and Rayleigh Waves by Surface Topographies", Report No. 79-05, Department of Civil Engineering, University of Southern California, Los Angeles, Calif., 1979.

32 Dravinski, M. "Scattering of SH Waves by Subsurface Topography". Journal of the Engineering Mechanics Division, ASCE, Vol. 108, No. EM1, Feb., 1982, pp. 1-7.

33 Dravinski, M. "Scattering of Elastic Waves by on Alluvial Valley". Journal of the Engineering Mechanics Division, ASCE, Vol. 108, No.EM1, Feb., 1982, pp. 19-31.

34 Wong, H.L. and Luco, J.E. "Dynamic Response of Rectangular Foundations to Obliquely Incident Seismic Waves", Earthquake Engineering and Structural Dynamics, Vol. 6, 1978, pp. 3–16.

35 Kobori, R., Minai, R. and Shinozaki, Y. "Vibrations of a Rigid Circular Disc on an Elastic Half-Space Subjected to Plane Waves", Theoretical and Applied Mechanics, 21, Univ. of Tokyo Press, 1973.

36 Luco, J.E. "Torsional Response of Structures to Obliquely Incident Seismic SH Waves", Earthquake Engineering and Structural Dynamics, Vol. 4, 1976, pp. 207–219.

37 Domínguez, J. "Response of Embedded Foundations to Travelling Waves". Research Report R78-24, Department of Civil Engineering, Massachusetts Institute of Technology, Cambridge, Mass., Aug., 1978.

38 Abascal, R. and Domínguez, J. "Dynamic behavior of strip footings on non-homogeneous viscoelastic soils", Proc. Int. Symposium on Dynamic Soil-Struct. Int., Minneapolis, Minnesota, 1984, Balkema, Roterdam.

39 Syrakos, C.C. "Dynamic Response of two Dimensional Foundations". Ph D. Th., Univ. Minnesota, Minneapolis, MN, 1984.

40 Karabalis, D.L. "Dynamic response of three dimensional foundations". Ph. D. Thesis, Univ. of Minnesota, Minneapolis, 1984.

41 Eringen, A.C. and Suhubi, E.S. "Elastodynamics", Academic Press, New York, 1975.

42 Wheeler, L.T. and Sternberg, E. "Some Theorems in Classical Elastodynamics", Archive for Rational Mechanics and Analysis, 1968, 31, 51.

43 Stokes, G.G. "On the Dynamical Theory of Diffraction". Transactions of the Cambridge Philosophical Society, 1984, 9, 1.

44 Graffi, D. "Sul Theorema di Reciprocitá nella Dinamica dei Corpo Elasticiti", Memorie della Accademia delle Scienze, 1946–7, 4 Series 10, 103.

45 Love, A.E.H. "The propagation of wave motion in an isotropie elastic solid medium". Proc. London Math. Soc., 2, 1, 1904, pp. 231–344.

46 Domínguez, J. and Abascal, R. "On fundamental solutions for the BIEM in static and dynamic elasticity", Eng. Analysis, 1, 1984, pp. 128–134.

47 Kupradze, V.D. "Dynamical Problems in Elasticity", Progress in Solid Mechanics (eds. Sneddon. I.N. and Hill, R.), Vol. 3, North-Holland, Amsterdam, 1963.

48 Cole, D.M., Kosloff, D.D. and Minster, J.B. "A Numerical Boundary Integral Equation Method for Elastodynamics I", Bulletin of the Seismological Society of America, Vol. 68, 1978, pp. 1331–1357.

49 Brebbia, C.A. "The BEM for engineers", Pentech Press, London, 1978.

50 Banerjee, P.K. and Butterfield, R. "Boundary Element Methods in Engineering Science", McGraw-Hill, London, 1981.

51 Kitahara, M. "Applications of BIEM to eigenvalue problems of elastodynamics and thin plates". Ph.D. Thesis Kyoto University, 1984.

52 Abascal, R, and Domínguez, J. "Sobre la extensión y tamaño de la malla de elementos de contorno en los problemas de interacción sueloestructura", Ier Congreso Iberoamericano de Met. Comp. en Ingeniería. Madrid, 1985.

53 Elsabee, F. and Morray, J.P. "Dynamic Behavior of Embedded Foundations", M.I.T. Research Report R77-33, September, 1977.

54 Kansel, E. and Ushijima, R. "Vertical and torsional stiffness of cylindrical footing", M.I.T., Report R-79-63, Civil Eng. Dept., 1979.

55 Gazetas, G.C. and Roësset, J.M. "Forced Vibrations of Strip Footings on Layered Soils", Methods of Structural Analysis, ASCE, Vol. 1, 1976, pp. 115–131.

56 Gazetas, G.C. and Roësset, J.M. "Vertical Vibrations of Machine Foundations", Journal of the Geotechnical Engineering Division, ASCE, Vol. 105, No. GT12, Dec., 1979, pp. 1435–1454.

57 Abascal, R. and Domínguez, J. "Vibrations of Footings on Zoned Viscoelastic Soils. I", Journal of Mechanical Engineering, ASCE. Accepted. To be published 1986.

58 Mayr, M. "Ein Integralgleichungsverfahren zur Lösung rotationssymmetrischer Elastizitätsprobleme", Dissertation, TU, München, 1975.

59 Kermanidis, T.A. "Numerical Solution for Axially Symmetrical Elasticity Problems", Journal of Solids and Structures, 1975, 11, 493.

60 Cruse, T.A., Snow, D.A. and Wilson, R.B. "Numerical Solutions in Axisymmetric Elasticity", Computer and Structures, 1977, 7, 445.

61 Rizzo, F.J. and Shippy, D.J. "Some observations on Kelvin's solution in classical elastostatics as a double tensor field with implications for Somigliana's integral", J. of Elast., 13, 1983, pp. 91–97.

62 Wilson, E. "Structural analysis of axisymmetric solids", AIAA Journ., 3, 12, 1965, p. 2269.

63 Cano Hurtado, J.J. "Cálculo de impedancias dinámicas de zapatas circulares rígidas en terrenos estratificados con amortiguamiento histerético". Tesis Doctoral, Univers. de Valencia, 1985.

64 Chapel, F. "Application de la méthode des équations intégrales à la dynamique des sols. Structures sur pieux". These presentee a l'ecole Central des Arts et Manufactures. 1981.

65 Luco, J.E. "LUCON: Theoretical and verification manual". Bechtel Power Corp., 1974.

66 Abascal, R. and Domínguez, J. "Dynamic response of embedded strip foundations subjected to obliquely incident waves". Proc. of the 7th Int. Conf. on BEM, Como, Italy, 1985.

67 Whittaker, W.L. and Christiano, P. "Dynamic response of flexible plates bearing on an elastic half-space", Proc. of ASCE, Jour. of Eng. Mech., 1, 1982, pp. 133–154.

Chapter 3

Boundary Integral Equation Methods for Consolidation Problems

by N. Nishimura

3.1 Introduction

The mechanical behavior of the soil is characterized by the presence of the fluid phase. Indeed, the interaction between the soil skeleton and fluid phase affects the behavior of soil in various situations. This interaction is especially important when the applied compression squeezes the fluid out of the soil. This phenomenon, or what was called 'consolidation' originally, was so typical as a subject in soil mechanics that the name 'consolidation analysis' has almost become a generic term for any soil deformation computations which take the effect of fluid into account. It is in this sense that we use the word 'consolidation' in this chapter.

The first attempt to analyze this phenomenon was due to Terzaghi [1]. He introduced the celebrated principle of effective stress and showed that in one dimensional cases the pore pressure is governed by the heat equation. Biot [2], among others, extended Terzaghi's ad hoc theory to the general three dimensional case, establishing a set of equations for poro-elasticity which is now called after his name. Biot's analysis made the structure of the problem clear, paving the way for further modification of models such as the use of the elastoplastic constitutive equation for soil skeleton, etc.

As in other equations in engineering, Biot's equation is difficult to solve analytically except in simple cases. Actually, most problems of practical interest require some numerical methods, such as FDM, FEM, etc. Although some FEM formulations (see e.g. Sandhu and Wilson [3]) are fairly well-accepted nowadays, they are still not without accuracy problems. For example, these methods have difficulty computing the initial and short time responses accurately. In addition, these numerical methods do not provide such information as how smooth the solution should be (as a matter of fact, consolidation solutions can suffer discontinuities), or what kind of singularity we are expecting, etc. Without these information, however, it would be difficult to implement a sufficiently accurate computer code. On the other hand, the boundary integral equation method is expected to be free of these drawbacks because it uses the exact representations of solutions. Moreover, the efficiency of the numerical code achieved by the 'boundary-only' formulation, although somewhat over-emphasized so far, still remains the characteristic feature of this method. These considerations motivated several researchers to investigate BIE formulations for consolidation problems.

This chapter presents several BIE formulations for consolidation problems thus proposed. In order to discuss and compare different formulations, however, it is necessary to set a definite viewpoint. This is true especially in an application oriented subject like ours. In this regard the present author found his own formulation (Nishimura [4]. Unfortunately, this paper has a lot of typographical errors.) useful as a basis for seeing other methods. Hence, we will spend a considerable portion of this chapter describing his formulation. Specifically, we start this chapter with a brief derivation of the governing equations. Green's formula and the fundamental solutions of these equations establish potential representations for the displacement and pore pressure in the usual manner. We then discuss some applications of these representations to problems of applied mathematical interest. In particular, we determine the singularity caused by the inconsistency of initial and boundary conditions, such as the inconsistency of 'drained' boundary condition with any non-zero initial pressure. We then proceed to a numerical analysis based on this formulation. A comparison of the numerical solution with the exact solution proves the accuracy of our method. This chapter concludes by comparing our method to others.

3.2 Statement of the Problem

There are two ways of obtaining the governing equations for poro-elasticity; the thermodynamical method originally used by Biot [2], and the different but equivalent practical approach customarily used by soil engineers. We will use the latter method.

The linear poro-elastic model for fluid-saturated soil uses the following laws:

1) *Equilibrium*

$$\operatorname{div} \tau + \rho F = 0, \tag{1}$$

where τ stands for the stress, ρ for the density of the soil, and F for the body force per unit mass.

2) *Terzaghi's Principle of Effective Stress*

$$\tau = \tau' - p\mathbf{1}, \tag{2}$$

where p is the pore pressure, $\mathbf{1}$ is the identity tensor, and τ' is the effective stress.

3) *Continuity of the Fluid*

$$\operatorname{div} \dot{u} + \operatorname{div} v + m\dot{p} = i, \tag{3}$$

where u, v, m, and i stand for the displacement of the skeleton, fluid velocity, compressibility of the fluid ($=0$ usually), and the rate of fluid injection per unit volume, respectively. Also, we have used the superposed dot for time differentiation.

4) *Constitutive Equation for Skeleton*

$$\tau' = C[\nabla u], \tag{4}$$

where C is the elasticity tensor (4th order) having the usual symmetry and positivity.

5) *Darcy's Law*

$$v = -K(\nabla p - \rho_f F), \tag{5}$$

where ρ_f is the density of the fluid, and K is the permeability tensor (2nd order positive definite and symmetric).

In addition to these principles, we assume that C, K, ρ, ρ_f and m are constant. These equations lead to the following governing equations:

$$\Delta^* u - \nabla p + \rho F = 0, \tag{6}$$

$$\operatorname{div} \dot{u} + m\dot{p} - K \cdot \nabla\nabla p = g, \tag{7}$$

where

$$\Delta^* u = \operatorname{div} C[\nabla u] \tag{8}$$

and

$$g = i - \rho_f K \cdot \nabla F. \tag{9}$$

When the soil is elastically isotropic, we have

$$\Delta^* u = \mu \Delta u + (\lambda + \mu) \nabla \operatorname{div} u, \tag{10}$$

where (λ, μ) are Lame's constants.

The following Green's identity holds for any sufficiently smooth pairs of vectors u, u^* and scalars p, p^*:

$$\int_{t_1}^{t_2} \int_{\partial D} [\dot{u}^* \cdot s - \dot{s}^* \cdot u - p^* n \cdot K\nabla p + n \cdot K\nabla p^* p] \, dS \, dt$$

$$= \int_{t_1}^{t_2} \int_{D} [\dot{u}^* \cdot (\Delta^* u - \nabla p) - (\Delta^* \dot{u}^* + \nabla \dot{p}^*) \cdot u$$

$$+ p^* (\operatorname{div} \dot{u} + m\dot{p} - K \cdot \nabla\nabla p) - (\operatorname{div} \dot{u}^* - m\dot{p}^* - K \cdot \nabla\nabla p^*)p] \, dV \, dt$$

$$- \int_{D} [p^* (\operatorname{div} u + mp)]_{t_1}^{t_2} \, dV, \tag{11}$$

where D is a domain (assumed to be bounded only for simplicity) with a smooth boundary ∂D having an outward unit normal vector n, and s and s^* are the surface traction and 'adjoint' surface traction defined by

$$s \overset{(a)}{=} (C[\nabla u] - 1p)n, \quad s^* \overset{(b)}{=} (C[\nabla u^*] + 1p^*)n. \tag{12a, b}$$

Equation (11) suggests to seek a solution of the following initial-boundary value problem:

Find a solution (u, p) of (6) and (7) in D and $t > 0$ subject to an initial condition

$$(\operatorname{div} u + mp)|_{t=0} = \theta \quad \text{in } D \tag{13}$$

and boundary conditions for $t > 0$

$$u = u_0 \quad \text{on } \partial D_u, \tag{14a}$$

$$s = s_0 \quad \text{on } \partial D_s, \tag{14b}$$

$$p = p_0 \qquad\qquad \text{on } \partial D_p, \tag{14c}$$

$$r := -\boldsymbol{n} \cdot \boldsymbol{K} \nabla p = r_0 \quad \text{on } \partial D_r, \tag{14d}$$

where θ, $\boldsymbol{u}_0$, s_0, p_0, and r_0 are given functions, and ∂D_u, ∂D_s, ∂D_p and ∂D_r stand for portions of the boundary ∂D which satisfy

$$\overline{\partial D_u \cup \partial D_s} \overset{(a)}{=} \partial D, \quad \partial D_u \cap \partial D_s \overset{(b)}{=} \varnothing, \tag{15a, b}$$

$$\overline{\partial D_p \cup \partial D_r} \overset{(c)}{=} \partial D, \quad \partial D_p \cap \partial D_r \overset{(d)}{=} \varnothing, \tag{15c, d}$$

with a superposed bar signifying the closure.

The reader may find no difficulty interpreting these statements in physical terms except for the initial conditions. This last condition is interpreted as follows: we first assume that some consolidation process was going on before $t = 0$. For example, we may assume the soil was at rest under no loading for $t < 0$. i.e., $\boldsymbol{u}(x, t) = \boldsymbol{0}$, $p(x, t) = 0$ for $t < 0$, $x \in D$. We then integrate Eq. (7) with respect to time from $-\Delta t$ to Δt to obtain

$$[\operatorname{div} \boldsymbol{u} + mp]_{t=\Delta t} - [\operatorname{div} \boldsymbol{u} + mp]_{t=-\Delta t}$$

$$= \int_{-\Delta t}^{\Delta t} (g + \boldsymbol{K} \cdot \nabla\nabla p)\, dt. \tag{16}$$

As $\Delta t \downarrow 0$, the right hand side tends to zero for any regular p and i. Hence we have

$$[\operatorname{div} \boldsymbol{u} + mp]_{t=0^+} = [\operatorname{div} \boldsymbol{u} + mp]_{t=0^-} := \theta. \tag{17}$$

In particular, $\theta = 0$ for the case of quiescent past (i.e. sudden loading).

Under certain conditions of regularity, we can use a standard argument to show that the solution of the above mentioned problem is unique, with the exceptions of the case $\partial D_u = \varnothing$ where the solution is determined to within a rigid motion, and the case $\partial D_p = \partial D_s = \varnothing$ and $m = 0$ where the pressure involves an undetermined additional constant. Also, we can show, by using Eq. (11), that the solvability conditions

$$\int_{\partial D} s(x, t)\, dS + \int_D \rho F(x, t)\, dV = \boldsymbol{0}; \quad t > 0 \tag{18}$$

and

$$\int_{\partial D} x \times s(x, t)\, dS + \int_D x \times \rho F(x, t)\, dV = \boldsymbol{0}; \quad t > 0 \tag{19}$$

must be satisfied when $\partial D_u = \varnothing$, while

$$\int_{\partial D} \boldsymbol{u} \cdot \boldsymbol{n}|_{t=t}\, dS + \int_0^t \int_{\partial D} r\, dS\, dt$$

$$= \int_0^t \int_D g\, dV\, dt + \int_D \theta\, dV; \quad t > 0 \tag{20}$$

is required if $m = 0$ and $\partial D_s = \partial D_p = \varnothing$.

3.3 Potential Representation of Solution

Let $\dot{U}$ (2nd order tensor), $\dot{V}$ (vector), P (vector), and Q (scalar) be functions which satisfy the 'causality' (i.e. $\dot{U}$ etc. vanish for $t < 0$) and the equation

$$
\begin{pmatrix} \varDelta^*, & -V\dfrac{\partial}{\partial t} \\[2ex] -V, & -m\dfrac{\partial}{\partial t} + K\cdot VV \end{pmatrix}
\begin{pmatrix} \dot{U} & \dot{V} \\ P & Q \end{pmatrix}
= -\begin{pmatrix} 1\delta(t)\delta(x), & 0 \\ 0, & \delta(t)\delta(x) \end{pmatrix}, \qquad (21)
$$

where $\delta(\,\cdot\,)$ is the Dirac delta. These functions are called the fundamental solutions of Eqs. (6) and (7). Since the functions $\dot{U}^*$, $\dot{V}^*$, P^*, and Q^* defined by

$$
\begin{pmatrix} \dot{U}^*(x,t), & \dot{V}^*(x,t) \\ P^*(x,t), & Q^*(x,t) \end{pmatrix}
= \begin{pmatrix} -\dot{U}(-x,-t), & -\dot{V}(-x,-t) \\ P(-x,-t), & Q(-x,-t) \end{pmatrix} \qquad (22)
$$

satisfy the 'anti-causality' (i.e. $\dot{U}^*$ etc. vanish for $t > 0$) and the equation

$$
\begin{pmatrix} \varDelta^*, & V\dfrac{\partial}{\partial t} \\[2ex] V, & -m\dfrac{\partial}{\partial t} - K\cdot VV \end{pmatrix}
\begin{pmatrix} \dot{U}^* & \dot{V}^* \\ P^* & Q^* \end{pmatrix}
= \begin{pmatrix} 1\delta(t)\delta(x), & 0 \\ 0, & \delta(t)\delta(x) \end{pmatrix}, \qquad (23)
$$

we may substitute $\dot{U}^*$ and P^* for $\dot{u}^*$ and p^* in Eq. (11) to obtain a potential representation for u, and manipulate similarly to find an expression for p.

There is a well-established general procedure for calculating the fundamental solution for linear PDEs with constant coefficients (See Dubois & Lachat [5]). As a matter of fact, the use of Fourier transform gives

$$
\dot{U}(x,t) = U_0(x)\delta(t) + \dot{U}(x,t)H(t), \qquad (24a)
$$

$$
\dot{V}(x,t) = V_0(x)\delta(t) + \dot{V}(x,t)H(t), \qquad (24b)
$$

where

$$
U_0 = F^{-1}\left(\varDelta^{*^{-1}}(\xi) - \frac{\varDelta^{*^{-1}}(\xi)\xi \otimes \varDelta^{*^{-1}}(\xi)\xi}{m + \xi\cdot\varDelta^{*^{-1}}(\xi)\xi} \right), \qquad (25a)
$$

$$
V_0 = F^{-1}\left(\frac{-i\varDelta^{*^{-1}}(\xi)\xi}{m + \xi\cdot\varDelta^{*^{-1}}(\xi)\xi} \right), \qquad (25b)
$$

$$
\dot{U} = F^{-1}\left(\varDelta^{*^{-1}}(\xi)\xi \otimes \varDelta^{*^{-1}}(\xi)\xi \frac{\xi\cdot K\xi\,\mathrm{ex}(\xi,t)}{(m + \xi\cdot\varDelta^{*^{-1}}(\xi)\xi)^2} \right), \qquad (25c)
$$

$$
\dot{V} = F^{-1}\left(i\varDelta^{*^{-1}}(\xi)\xi \frac{\xi\cdot K\xi\,\mathrm{ex}(\xi,t)}{(m + \xi\cdot\varDelta^{*^{-1}}(\xi)\xi)^2} \right), \qquad (25d)
$$

and

$$
\mathrm{ex}(\xi,t) = e^{-\xi\cdot K\xi t/(m+\xi\cdot\varDelta^{*^{-1}}(\xi)\xi)}. \qquad (26)
$$

In these expressions, we have used F^{-1} for Fourier inverse transform defined by

$$F^{-1} \cdot = \frac{1}{(2\pi)^N} \int_{R^N} \cdot \, e^{i\xi \cdot x} \, d\xi \tag{27}$$

for N dimensional problems, $\varDelta^{*^{-1}}(\xi)$ for the matrix inverse to the one obtained by replacing V by ξ in $\varDelta^*$, and $H(t)$ for Heaviside's step function. Also, we have

$$P = H(t)F^{-1}\left(\frac{-i\varDelta^{*^{-1}}(\xi)\xi}{m + \xi \cdot \varDelta^{*^{-1}}(\xi)\xi} \, ex(\xi, t)\right), \tag{28a}$$

$$Q = H(t)F^{-1}\left(\frac{ex(\xi, t)}{m + \xi \cdot \varDelta^{*^{-1}}(\xi)\xi}\right). \tag{28b}$$

With these preliminaries we can write down the potential representations of u and p as follows:

$$\begin{aligned}
\tilde{u}(x, t) = &\int_{\partial D} U_0(x - y)s(y, t) \, dS - \int_{\partial D} S_0(x, y)u(y, t) \, dS + \int_D U_0(x - y)\rho F(y, t) \, dV \\
&+ \int_{\partial D} \int_0^t \dot{U}(x - y, t - s)s(y, s) \, ds \, dS - \int_{\partial D} \int_0^t \dot{S}(x, y, t - s)u(y, s) \, ds \, dS \\
&- \int_{\partial D} \int_0^t P(x - y, t - s)r(y, s) \, ds \, dS + \int_{\partial D} \int_0^t R(x, y, t - s)p(y, s) \, ds \, dS \\
&+ \int_D \int_0^t \dot{U}(x - y, t - s)\rho F(y, s) \, ds \, dV + \int_D \int_0^t P(x - y, t - s)g(y, s) \, ds \, dV \\
&+ \int_D P(x - y, t)\theta(y) \, dV, \tag{29}
\end{aligned}$$

and

$$\begin{aligned}
\tilde{p}(x, t) = &\int_{\partial D} V_0(x - y) \cdot s(y, t) \, dS - \int_{\partial D} T_0(x, y) \cdot u(y, t) \, dS \\
&+ \int_D V_0(x - y) \cdot \rho F(y, t) \, dV + \int_{\partial D} \int_0^t \dot{V}(x - y, t - s) \cdot s(y, s) \, ds \, dS \\
&- \int_{\partial D} \int_0^t \dot{T}(x, y, t - s) \cdot u(y, s) \, ds \, dS - \int_{\partial D} \int_0^t Q(x - y, t - s)r(y, s) \, ds \, dS \\
&+ \int_{\partial D} \int_0^t W(x, y, t - s)p(y, s) \, ds \, dS + \int_D \int_0^t \dot{V}(x - y, t - s) \cdot \rho F(y, s) \, ds \, dV \\
&+ \int_D \int_0^t Q(x - y, t - s)g(y, s) \, ds \, dV + \int_D Q(x - y, t)\theta(y) \, dV. \tag{30}
\end{aligned}$$

In these formulas, u and p stand for the functions defined by

$$\tilde{u}, \tilde{p} = \begin{cases} u, p & \text{in } D \\ 0 & \text{in } R^N \backslash D. \end{cases} \tag{31}$$

Also, $S_0, \dot{S}, R, T_0, \dot{T}$ and W are certain kernel functions whose definitions may be inferred readily from Eq. (11). Later we shall present the explicit forms of these functions for isotropic cases.

3.4 Analytical Applications

This section will show some analytical applications of the foregoing analysis to the investigation of the behavior of the solutions u and p. We shall state two results concerning the initial values of the solutions with sketches of proofs. These results will be used later to implement an accurate numerical method of solution.

By the word 'initial value' we mean

$$\lim_{t \downarrow 0} u(x,t), \quad x \in \bar{D}. \tag{32}$$

We have the following results:

(i) Let u_i be the solution of the equation

$$\mathrm{div}\left(C + \frac{1 \otimes 1}{m}\right)[\nabla u_i] + \rho F - \frac{1}{m} \nabla \theta = 0 \quad \text{in } D, \tag{33}$$

subject to the boundary conditions

$$u_i(x) = u_0(x,0); \quad x \in \partial D_u, \tag{34a}$$

$$\left(C + \frac{1 \otimes 1}{m}\right)[\nabla u_i(x)]n = s_0(x,0) - n\theta(x)/m; \quad x \in \partial D_s. \tag{34b}$$

Then $u(x,0)$ and $p(x,0)$ for non-zero m satisfy

$$u(x,0) = u_i(x), \tag{35a}$$

$$p(x,0) = (\theta(x) - \mathrm{div}\, u_i(x))/m; \quad x \in D. \tag{35b}$$

In other words, the initial values of u and p in D for $m \neq 0$ coincide with the displacement and $(\theta - \text{volume strain})/m$ of the elastostatic BVP with $C + 1 \otimes 1/m$ as the elasticity tensor, $u_0(x,0)$ as the boundary displacement on ∂D_u, $s_0(x,0) - \theta(x)n(x)/m$ as the surface traction on ∂D_s, and $\rho F(x,0) - \nabla \theta(x)/m$ as the body force. For $m = 0$, we introduce a pair (u_i, p_i) which satisfies

$$\Delta^* u_i - \nabla p_i + \rho F = 0, \tag{36a}$$

$$\mathrm{div}\, u_i = \theta \quad \text{in } D \tag{36b}$$

and

$$u_i(x,0) = u_0(x,0) \quad \text{on } \partial D_u, \tag{37a}$$

$$C[\nabla u_i(x)]n - p_i(x)n = s_0(x,0) \quad \text{on } \partial D_s. \tag{37b}$$

Then we have

$$u(x,0) = u_i(x), \tag{38a}$$

$$p(x,0) = p_i(x) \quad x \in D. \tag{38b}$$

In mechanical terms, we may say that the initial values of u and p in D for $m = 0$ coincide with the displacement and the indeterminate pressure of the elastostatic BVP with the original elasticity tensor, $u_0(x,0)$, $s_0(x,0)$ and $\rho F(x,0)$ as given data, and div $u = \theta$ (in D) as a constraint.

As a corollary, we have the following:

When any of the data (u_0, s_0, F) jumps as in the case of sudden change of loading, the displacement u and pressure p in D also jump. The amount of this jump is read off from the foregoing statements for initial values by replacing u, p etc. for $t = 0$ by the jumps of u, p etc., respectively (θ is replaced by 0).

Here, we shall give a sketch of an independent proof of the corollary for non-zero m. Equation (29) shows that the jump of u, denoted by $[[u]]$, is written as

$$[[\bar{u}(x,t)]] = \int_{\partial D} U_0(x - y)[[s(y,t)]]\, dS - \int_{\partial D} S_0(x - y)[[u(y,t)]]\, dS$$
$$+ \int_{D} U_0(x - y)\rho[[F(y,t)]]\, dV. \tag{39}$$

On the other hand, Eq. (25a) shows that U_0 is the fundamental solution of elasticity with $C + 1 \otimes 1/m$ as the elasticity tensor. Similarly, S_0 is seen to be the corresponding double layer kernel. Hence $[[u]]$ has the above mentioned interpretation. In the same manner we obtain the similar result for p. This concludes the proof.

Actually, all these statements are equivalent to the fact that

$$[[\operatorname{div} u + mp]] = 0. \tag{40}$$

We have already used this relation implicitly in discussing the interpretation of θ. Once one accepts this fact, all the foregoing results follow immediately. For example, we have

$$\operatorname{div}\left[\left[C[\nabla u] + \frac{1}{m}\operatorname{div} u\right]\right] + [[\rho F]] = 0, \tag{41}$$

where we have used Eqs. (6) and (40). This formula is equivalent to the statement of the corollary.

The result (i) shows that the initial field is determined only by the initial values of u_0, s_0, F, and θ. In particular, the initial pressure p and, hence, the limit

$$\lim_{x \to x_0} \lim_{t \downarrow 0} p(x,t); \quad x \in D, \quad x_0 \in \partial D \tag{42}$$

are determined from these data. However, for $x_0 \in \partial D_p$, we would have to prescribe p including its limit as $t \downarrow 0$, i.e.,

$$\lim_{t \downarrow 0} \lim_{x \to x_0} p(x,t); \quad x \in D, \quad x_0 \in \partial D. \tag{43}$$

There is no guarantee that these limits coincide because we have no way of knowing the former limit beforehand without undertaking any computation. Also on ∂D_r, we will have to evaluate the latter (from the former) in order to carry out a numerical BIEM using a linear (or higher order) time interpolation. For example the 7th integral in Eq. (29) uses this limit as the initial value of p on ∂D_r. Hence we are led to the following question: what is the relation between these two limits? The formulation in the last section gives the answer without difficulty. Actually, it says:

(ii) Assume the limits in Eqs. (42) and (43) are finite. Also assume that $\partial p/\partial n$ may have a singularity of a power form in t for $t \approx 0$, $x \in \partial D$. Then we have

$$\lim_{x \to x_0} \lim_{t \downarrow 0} p(x, t) = \lim_{t \downarrow 0} \lim_{x \to x_0} p(x, t) - \sqrt{\frac{\pi n \cdot Kn}{m + n \cdot \Delta^{*^{-1}}(n)n}} \gamma(x_0);$$

$$x \in D, \, x_0 \in \partial D, \qquad (44)$$

where

$$\gamma(x_0) = \lim_{t \downarrow 0} \frac{\partial p}{\partial n}(x_0, t)\sqrt{t}. \qquad (45)$$

To see this, one may carry out a direct limit calculation based on Eq. (30) and the result of (i). The detail of this calculation, however, is too technical to be included here. As a special case, we see that the fluid velocity $(\partial p/\partial n)$ on ∂D_r has a singularity of order $1/\sqrt{t}$ near $t = 0$ if the two limits in Eqs. (42) and (43) are both finite and different. Also, if the singularity of $\partial p/\partial n$ is weaker than $O(1/\sqrt{t})$ near $t = 0$ at a point on the boundary, the two limits in Eqs. (42) and (43) are identical there. Specifically, on the undrained boundary $(r = 0)$ these two limits are equal. Therefore we may compute the limit in Eq. (42) considering (i) and then use the same value for the initial value of the pressure on the boundary [Eq. (43)]. This result justifies the BIEM formulation which will be presented in the next section.

3.5 Numerical Method

This section discusses the numerical methods based on the formulation in 3.3. We shall, however, restrict ourselves to 2-dimensional isotropic case neglecting the compressibility of the fluid. These assumptions are equivalent to

$$C[\nabla u] = \lambda 1 \operatorname{div} u + \mu(\nabla u + (\nabla u)^T), \qquad (46a)$$

$$K = k1 \qquad (46b)$$

and

$$m = 0, \qquad (47)$$

where k is the permeability constant. This simplification makes the explicit forms of the fundamental solutions available (although $m = 0$ is not essential for the availability of the fundamental solutions). Also, we shall assume that F, i and θ vanish identically. In physical terms, this amounts to considering a response to a sudden loading under no body force and fluid injection. With these assumptions, we obtain the following set of integral equations:

$$\frac{1}{2}u(x, t) = \int_{\partial D} U_0(x - y)s(y, t)\,dS - \int_{\partial D} S_0(x, y)u(y, t)\,dS - \oint_{\partial D} P_0(x - y)q(y, t)\,dS$$

$$+ \int_{\partial D} \int_0^t \dot{U}(x - y, t - s)s(y, s)\,ds\,dS - \oint_{\partial D} \int_0^t \dot{S}(x, y, t - s)u(y, s)\,ds\,dS$$

$$- \int_{\partial D} \int_0^t \dot{P}(x - y, t - s)q(y, s)\,ds\,dS + \int_{\partial D} \int_0^t R(x, y, t - s)p(y, s)\,ds\,dS \quad (48)$$

and

$$\tfrac{1}{2}p(x,t) = \oint_{\partial D} V_0(x-y) \cdot s(y,t)\,dS - \text{pf} \int_{\partial D} T_0(x,y) \cdot u(y,t)\,dS$$

$$+ \oint_{\partial D} \int_0^t \dot{V}(x-y,t-s) \cdot s(y,t)\,ds\,dS$$

$$- \text{pf} \oint_{\partial D} \int_0^t \dot{T}(x,y,t-s) \cdot u(y,s)\,ds\,dS$$

$$- \int_{\partial D} \int_0^t \dot{Q}(x-y,t-s)q(y,s)\,ds\,dS$$

$$+ \oint_{\partial D} \int_0^t W(x,y,t-s)p(y,s)\,ds\,dS \tag{49}$$

where

$$q(x,t) = \int_0^t r(x,s)\,ds, \tag{50a}$$

$$U_0(x-y) = -\frac{1}{4\pi\mu}(1\log R - \nabla R \otimes \nabla R), \tag{50b}$$

$$S_0(x,y) = \frac{1}{\pi R}(\nabla R \cdot n_y)\nabla R \otimes \nabla R, \tag{50c}$$

$$P_0(x-y) = V_0(x-y) = \frac{\nabla R}{2\pi R}, \tag{50d}$$

$$\dot{U}(x-y,t) = \frac{k}{\pi}\left[1\left(\frac{1-\exp}{2R^2}\right) - \nabla R \otimes \nabla R\left(\frac{1-\exp}{R^2} - \frac{\exp}{4C_v t}\right)\right], \tag{50e}$$

$$\dot{S}(x,y,t) = -\frac{2\mu k}{\pi R}\left\{[(1 - 4\nabla R \otimes \nabla R)(\nabla R \cdot n_y) + n_y \otimes \nabla R + \nabla R \otimes n_y]\right.$$
$$\left. \cdot \left(\frac{\exp}{4C_v t} - \frac{1-\exp}{R^2}\right) + (\nabla R \otimes n_y - \nabla R \otimes \nabla R(\nabla R \cdot n_y)\frac{R^2\exp}{8C_v^2 t^2}\right\}, \tag{50f}$$

$$\dot{P}(x-y,t) = -\frac{R\exp}{8\pi C_v t^2}\nabla R, \tag{50g}$$

$$R(x,y,t) = -\frac{k}{\pi}\frac{\partial}{\partial n_y}\cdot\left[\left(\frac{1-\exp}{2R}\right)\nabla R\right], \tag{50h}$$

$$T_0(x,y) = -\mu\frac{n_y - 2\nabla R(\nabla R \cdot n_y)}{\pi R^2}, \tag{50i}$$

$$\dot{V}(x-y,t) = -\frac{R\exp}{8\pi C_v t^2}\nabla R, \tag{50j}$$

$$\dot{T}(x,y,t) = \frac{\mu}{8\pi C_v}\left[\left(n_y - \nabla R(\nabla R \cdot n_y)\right)\frac{R^2\exp}{C_v t^3} - n_y\frac{2\exp}{t^2}\right], \tag{50k}$$

$$\dot{Q}(x - y, t) = \frac{1}{4\pi k}\left(\frac{R^2 \exp}{4C_v t^3} - \frac{\exp}{t^2}\right), \tag{50l}$$

$$W(x, y, t) = -\frac{k}{\pi}\frac{\partial}{\partial n_y}\left(\frac{\exp}{4kt}\right), \tag{50m}$$

with

$$C_v = k(\lambda + 2\mu), \tag{51a}$$

$$R = |x - y|, \tag{51b}$$

$$\exp = e^{-R^2/4C_v t}. \tag{51c}$$

Also, we have used the symbols $\displaystyle\oint$ for integral in the sense of Cauchy's principal value, pf for the finite part of integral, and V for the derivative with respect to x. In view of the possible singularity of r on ∂D discussed in the last section, we have replaced r by q using the integration by part on t. It is noted that these integral equations are totally free of volume integrals.

The numerical procedure goes as follows:

1. Determine the initial field. In the present context, this is done by using the BIEM for incompressible elasticity discussed by Kobayashi and Nishimura [6]. Unless the domain D has a hole the conventional BIE works with Poisson's ratio set equal to 0.5. If D does have a hole, one has to modify the kernel function. Otherwise, the resulting integral equation loses the uniqueness of the solution in spite of the uniqueness of the solution of the BVP itself. In this respect, the comment of Banerjee and Butterfield [7] claiming that the conventional BIEM works with v (Poisson's ratio) $= \frac{1}{2}$ is incorrect.

2. Now assume that all the boundary quantities for $t = 0, \Delta t, \ldots, (n - 1)\,\Delta t$ are known, where n is an integer and Δt is a time increment. The conventional BIEM formulation with some time interpolation reduces the integral equations (48) and (49) into their discretized counterparts. By solving these equations we can determine all the boundary quantities up to $t = n\Delta t$. The time integration can be done analytically at least with linear time interpolation.

3. u and p in D are determined by using Eqs. (48) and (49) with $(u/2, p/2)$ replaced by (u, p) and $\displaystyle\oint$ and pf taken as ordinary integrals.

The only possible numerical difficulty for this formulation is the evaluation of singular integrals. Especially, the computation of pf-integrals appears to be a formidable task. That this is not to be the case is shown below. To fix the idea, we use a linear time interpolation. The most singular integrals in Eq. (48) are those involving S_0 and $\dot{S}$. Combining these integrals we have

$$\int_{\partial D}\left[S_0(x, y) + \int_{t-\Delta t}^{t} \dot{S}(x, y, t - s)\left(1 - \frac{t - s}{\Delta t}\right)ds\right]\Omega(y)\,dS \tag{52}$$

as a typical form of the most singular integral, where $\Omega(y)$ is a certain shape function on ∂D. The kernel function is readily shown to behave as

$$S_0(x,y) + \int_{t-\Delta t}^{t} \dot{S}(x,y,t-s)\left(1 - \frac{t-s}{\Delta t}\right)ds$$

$$= \frac{1}{2\pi(\lambda + 2\mu)R}\left[(\mu \mathbf{1} + 2(\lambda + \mu)\nabla R \otimes \nabla R)(\nabla R \cdot n_y)\right.$$

$$\left. + \mu(n_y \otimes \nabla R - \nabla R \otimes n_y)\right] + O(R\log R), \tag{53}$$

where we have used Eqs. (50c) and (50f). Note that the $O(1/R)$ term in this expression is exactly the same as the double layer kernel of elastostatics. Similarly, we can show that U_0 and $\dot{U}$ give a kernel having the same singularity as the simple layer kernel of elastostatics. The kernels in Eq. (49) reduce their singularities much more dramatically than in Eq. (48). For example, a direct integration using Eqs. (50i) and (50k) shows

$$T_0(x,y) + \int_{t-\Delta t}^{t} \dot{T}(x,y,t-s)\left(1 - \frac{t-s}{\Delta t}\right)ds$$

$$= -\frac{\mu}{2\pi C_v \Delta t} n_y \log R + O(1), \tag{54}$$

thus removing a possible $1/R^2$ singularity completely. Summing up, we see that the evaluation of the integrals in Eqs. (48) and (49) does not require anything more than the integrations of elastostatic kernels plus some bounded functions. Furthermore, the coding for the present formulation should not be too difficult because one can do so just by modifying some existing BIEM program for elastostatics. This is because the singularities of the kernels of these problems are exactly one and the same.

Finally, we have one remark concerning the initial values on ∂D. When we set up discretized integral equations for $t = n\Delta t$, we evaluate time-boundary integrals in Eqs. (48) and (49) by using an interpolation of quantities such as u, p, etc. for $t = 0, \Delta t, \ldots, n\Delta t$. In this case what we mean by the quantities for $t = 0$ on ∂D are the limits of the form given in Eq. (43). Therefore, specifically on ∂D_r, we obtain what we need, or the limit in Eq. (43), by computing the limit in Eq. (42) from the initial field, followed by the use of Eq. (44).

3.6 Numerical Examples

In this section we show the applicability of the present formulation by simple examples.

3.6.1 Circular Disc

We consider a circular domain D having a radius of a. We solved the following initial- boundary value problem:

$$\text{Initial condition} \qquad \theta = 0, \tag{55a}$$

$$\text{Boundary condition} \qquad s \overset{(b)}{=} -p_0 n, \quad p \overset{(c)}{=} 0, \tag{55b, c}$$

$$\text{Poisson's ratio} \qquad \nu = 0, 1/3. \tag{55d}$$

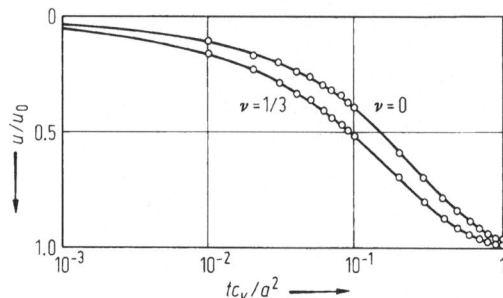

Fig. 1. Boundary displacement of a circular hole. Symbols: BIE, lines: exact, u_0: final displacement

For this analysis we used 32 linear isoparametric boundary elements. The time shape function is also chosen to be linear. All the integrals with respect to time are carried out analytically, whereas the spatial integrals are evaluated by use of the Gaussian cubature. As in elastostatics, we calculated principal value integrals in the integral equations by using the fact that the rigid body motion is a solution of our problem. The time increment was chosen as $\Delta t C_v/a^2 = \frac{1}{100}$ and was kept constant. The BIE results shown in Fig. 1 by symbols agree with the analytical solutions [8] shown by lines. The CPU time was about 3 sec. per step by using FACOM VP100 of Data Processing Center of Kyoto University.

3.6.2 Circular Hole in an Infinite Plane

We next consider a circular hole (radius $= a$) in an infinite plane. We assume that the infinite plane has an initial stress τ^0 and a vanishing initial pressure. We then 'excavate' a hole in a way that it satisfies the following conditions:

$$\text{Initial condition} \qquad \theta = 0, \tag{56a}$$

$$\text{Boundary condition} \quad s \overset{(b)}{=} 0, \quad p \overset{(c)}{=} 0. \tag{56b, c}$$

In the present analysis we set $\tau^0_{11} = -0.4\, p_0$, $\tau^0_{22} = -0.8\, p_0$, $\tau^0_{12} = 0$ and $\nu = 0$. Also, we have used 32 piecewise linear boundary elements of equal length. Fig. 2 shows the deformation of the boundary calculated by using the present formulation. The out-most circle represents the undeformed shape of the boundary and the least flat (flattest) curve shows the initial (final) shape. In this figure we have plotted two series of numerical results together, i.e., the displacements on ∂D for $0 < C_v t/a^2 < 1$ obtained with $C_v \Delta t/a^2 = \frac{1}{100}$ and those for $1 < C_v t/a^2 < 10$ with $C_v \Delta t/a^2 = \frac{1}{10}$. As it turned out this problem is particularly sensitive to the accuracy of the numerical integration. We therefore improved the accuracy of some of logarithmically singular integrals by using the fact that $(\boldsymbol{u} = \varGamma 1/|\boldsymbol{x}|, p = 0)$ and $(\boldsymbol{u} = \text{uniform shear}, p = 0)$ are solutions of Eqs. (6) and (7).

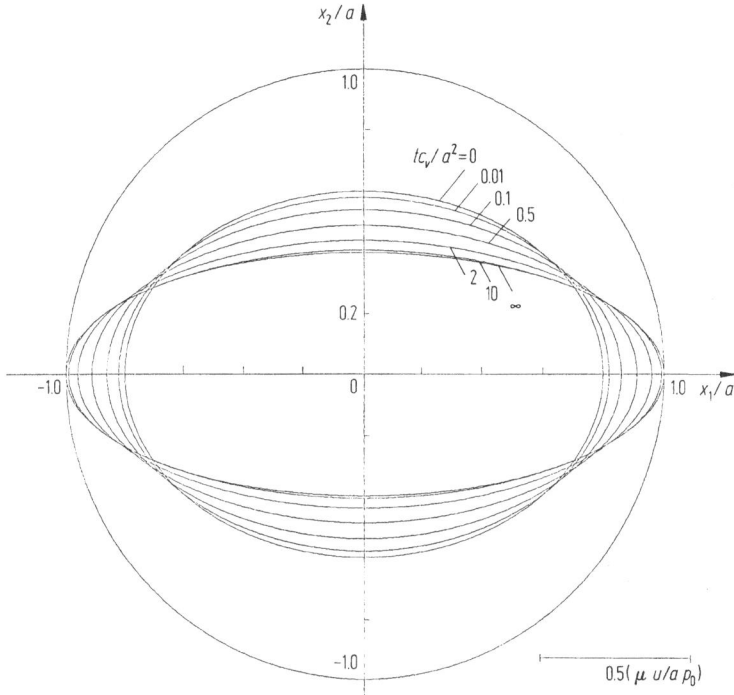

Fig. 2. Deformation of a circular hole subject to biaxial compression. Poisson's ratio = 0

3.6.3 Loaded Halfplane

As the third example we consider a loaded halfplane. The conditions are

Initial condition	$\theta = 0$,	(57a)
Boundary condition	$p = 0$, s: as given in the inset of Fig. 3,	(57b)
Poisson's ratio	$v = 0$.	(57c)

From a physical point of view the displacement field for this problem is rather pathological because it behaves logarithmically at the point of infinity. In addition the solution is not unique in the sense that one may superpose a rigid motion on one solution of this problem to obtain another. In civil engineering practice, however, one usually sets vertical displacements at certain points (usually taken away from the loaded portion) equal to zero superposing a rigid motion. We here follow this practice by setting $u_2 = 0$ at $x_1 = \pm 7a$, $x_2 = 0$. Hence the reader should remember that the pattern of the displacement, rather than its absolute value, bears physical meaning. Another peculiarity of this problem arises from the fact that the length of the boundary is infinite. Since our scheme cannot deal with such a

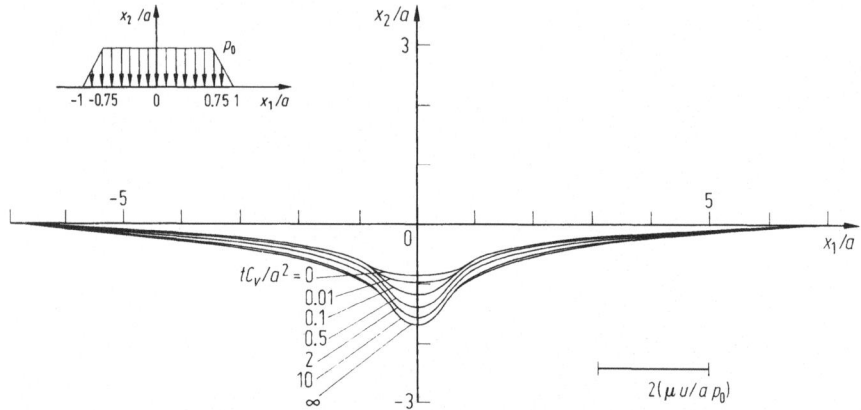

Fig. 3. Deformation of a loaded halfplane. Poisson's ratio $= 0$

boundary, we truncate it and take only the finite part of it into account in numerical analysis. In order to see the effect of this truncation we have carried out two numerical analyses, one with the truncation at $x_1 = \pm 9a$ and the other at $x_1 = \pm 20a$. This comparison has shown practically no difference between these two runs, thus justifying the method of truncation. The deformation pattern shown in Fig. 3 is obtained by using $C_v \Delta t/a^2 = \frac{1}{10}$ and $\frac{1}{100}$ ($\frac{1}{100}$ for $C_v t/a^2 < 1$ and $\frac{1}{10}$ for $C_v t/a^2 > 1$). Poisson's ratio is 0, and the number of (linear) boundary elements is 45. The boundary is truncated at $x_1 = \pm 20a$. The upper-most curve shows the initial deformation. As the time proceeds the deformation at the mid point of the boundary increases apparently.

3.7 Other Formulations

1. Formulation by Predeleanu

In Predeleanu [9] we can find the first attempt to formulate a direct BIEM for consolidation problems. To see this formulation we start with another form of Green's formula:

$$
\int_{t_1}^{t_2} \int_{\partial D} [u^* \cdot \dot{s} - s^* \cdot \dot{u} + p^* n \cdot K \nabla p^* - n \cdot K \nabla p^* p] \, dS \, dt
$$

$$
= \int_{t_1}^{t_2} \int_D [u^* \cdot (\Delta^* \dot{u} - \nabla \dot{p}) - (\Delta^* u^* + \nabla p^*) \cdot \dot{u}
$$

$$
- p^*(\operatorname{div} \dot{u} + m\dot{p} - K \cdot \nabla \nabla p) + (\operatorname{div} \dot{u}^* - m\dot{p}^* - K \cdot \nabla \nabla p^*)p] \, dV \, dt
$$

$$
- \int_D [(\operatorname{div} u^* - mp^*)p]_{t_1}^{t_2} \, dV. \tag{58}
$$

This formula is proven in the same manner as in Eq. (11). It would be easy to see that the substitution of appropriate fundamental solutions into (u^*, p^*) in Eq. (58) yields a set of potential representations for $(\dot{u}, p)$, reproducing the formulas of Predeleanu. Note that the last integral in Eq. (58) will pick up p as the initial condition.

The peculiarity of his formulation, in contrast to the present author's, is that it computes $\dot{u}$. This can, however, be a potential source of inconvenience since the solutions for consolidation problems may have jump discontinuities corresponding to a sudden change of loading. Namely, $\dot{u}(t)$ may behave like $\delta(t)$ making the numerical analysis including $\dot{u}$ difficult. In addition, his formulation is not free of volume integrals for the case of sudden initial loading. As a matter of fact his formulation requires p as the initial condition, and p does not necessarily vanish initially. The present author doubts that there could be feasible situations where p is an adequate choice for an initial condition.

2. Formulations of Banerjee and Butterfield; Kuroki et al.; and Garcia-Suarez and Alarcon

Banerjee and Butterfield [7] proposed both direct and indirect BIE formulations for consolidation with $m = 0$ utilizing BIEs for elastostatics and heat equation with inhomogeneity. Almost at the same time, Kuroki et al. [10] obtained essentially the same formulation (but with non-zero m) independently. The formulation of these authors uses integral representations for u and p derived separately from Eqs. (6) and (7), respectively. In the sequel we reproduce their direct formulation using our notation and sign convention which are different from theirs. We also use an assumption that F, g and m vanish just for simplicity.

They first regard Eq. (6) as the equation of elastostatics with $-\nabla p$ as an inhomogeneity (or, body force in the language of mechanics), and then use BIE formulations for elastostatics and divergence theorem to obtain an expression

$$u(x) = \int_{\partial D} \Gamma(x - y)s(y)\,dS - \int_{\partial D} \Gamma_l(x, y)u(y)\,dS$$
$$+ \int_D \operatorname{div}_y \Gamma(x - y)p(y)\,dV, \tag{59}$$

where $\Gamma(x)$ is the fundamental solution of elastostatics:

$$\Delta^* \Gamma(x) = -1\delta(x), \tag{60}$$

and $\Gamma_l(x, y)$ is the double layer kernel. In order to derive an expression for p, they rewrite Eq. (7) as

$$A\dot{p} - K \cdot \nabla\nabla p + (\operatorname{div} \dot{u} - A\dot{p}) = 0, \tag{61}$$

where A is an arbitrary constant. Taking the expression in parentheses as an inhomogeneity, they obtain the following representation for p:

$$A\tilde{p}(x,t) = \int\limits_{\partial D} \int\limits_0^t G(x-y,t-s)n(y)\cdot KVp(y,s)\,ds\,dS$$

$$- \int\limits_{\partial D} \int\limits_0^t n(y)\cdot KV_y G(x-y,t-s)p(y,s)\,ds\,dS$$

$$- \int\limits_D \int\limits_0^t G(x-y,t-s)(\operatorname{div}\dot{u}(y,s) - A\dot{p}(y,s))\,ds\,dV$$

$$+ A \int\limits_D G(x-y,t)p(y,0)\,dV, \tag{62}$$

where G is the fundamental solution of heat equation:

$$\left(\frac{\partial}{\partial t} - \frac{1}{A}K\cdot VV\right)G(x,t) = \delta(t)\delta(x). \tag{63}$$

For numerical analysis, they use time-marching computations essentially analogous to what we have seen in Sect. 3.5. Specifically, they solve an integral equation
obtained from Eq. (62) successively giving a small time increment for each time step.
They give p as the initial condition and evaluate the time derivatives in the time-
volume integral in Eq. (62) by using backward difference formulas. Obviously, their
formulation is not of boundary type and it requires the computation of div u and p
at a sufficient number of points in D. For time integration, Banerjee and Butterfield
proposed two methods. The first method uses Eq. (62) as it is, i.e., it computes the
time integral in Eq. (62) from 0 to t for determining p for $t = t$. The second method
employs Eq. (62) with $t = \Delta t$, and subsequently shifts the origin of time, taking the
results of the previous time step to be the new initial condition. Kuroki et al. use
the second method. Surprisingly, these authors choose the same value for A (for
$m = 0$). They put

$$A = \frac{1}{3\lambda + 2\mu} \tag{64}$$

for isotropic case. This choice makes the inhomogeneity in Eq. (61) proportional
to tr $\dot{\tau}$.

The present author thinks that their formulation is useful when it uses the
second method of time integration of Banerjee and Butterfield. This keeps the
amount of computation for each step almost equal. On the other hand the first
method would make the computation increasingly burdensome as the time grows.
In the formulation of the present author, however, the first method is worthwhile because the second method is possible only with the loss of 'boundary-only'
property

One of the advantages of their formulation over the present author's is that
theirs is based on simpler equations. It uses a time-boundary integral equation only
for a scalar p, and computes other quantities (u, etc.) using potential representations
including only spatial integrals such as Eq. (59). On the other hand, the formulation
of the present author uses complicated formulas like Eq. (29) for all the field
quantities. However, the B-B-K formulation may not yield very accurate results

when the loading changes rapidly. In addition, their formulation relies on a ques-
tionable trick of using div $\dot{u} - A\dot{p}$ of the previous time step for calculating the
time-volume integral in Eq. (62). (Does not it follow that one can solve any
equation in this manner?!) Also, p is awkward as the initial condition as has been
pointed out. As a matter of fact, it is not difficult, at least theoretically, to get
rid of part of this awkwardness of their formulation. Indeed, we can transform Eq.
(62) into

$$\text{div } \tilde{u}(x, t) = \int_{\partial D} \int_0^t G(x - y, t - s)n(y) \cdot K\nabla p(y, s) \, ds \, dS$$

$$- \int_{\partial D} \int_0^t n(y) \cdot K\nabla_y G(x - y, t - s)p(y, s) \, ds \, dS$$

$$- \int_D \int_0^t \frac{d}{dt} G(x - y, t - s)(\text{div } u(y, s) - Ap(y, s)) \, ds \, dV$$

$$+ \int_D G(x - y, t)\text{div } u(y, 0) \, dV. \tag{65}$$

This rewriting eliminates both the awkwardness of initial condition and the time
differentiation in the time-volume integral. For numerical solution of Eq. (65) one
may, by and large, follow the methods of Banerjee and Butterfield by using div u
etc. of the previous time step for evaluating volume integrals. After determining div u
in D by using Eq. (63), one may calculate u and p, by utilizing the BIE for elastostatics
with given volume strain, or what is the same thing, the BIE for incompressible
elastostatics with inhomogeneity (see Kobayashi & Nishimura [6]). Also, the present
author thinks that it is not necessary to choose A as given in Eq. (64). One may, for
example, choose A so that the fundamental solution G has the 'correct' exponential
term $\exp(-R^2/4C_v t)$ inferred from Eq. (50). This gives

$$A = \frac{1}{\lambda + 2\mu}. \tag{66}$$

At any rate, we may say that the B-B-K formulation can be practical because
of its simplicity although for analyses requiring high accuracy it is advisable to use
the formulation of present author.

Garcia-Suarez and Alarcon [11] also proposed a BIE formulation similar to
the one discussed above. The only difference is that Garcia-Suarez and Alarcon
used a BIEM for a Helmholtz type equation after transforming eq. (61) into a
Helmholtz type equation with inhomogeneity by using a backward difference to $\dot{p}$.

In [12, 13] Aramaki et al. extended the formulation of Kuroki et al. to the case
of a soil having a thin layer of high or low permeability.

3.8 Concluding Remarks

1. The fundamental solutions for 3-dimensional isotropic case are also available.
For $m = 0$, they take the following forms:

$$
\begin{pmatrix} \dot{\mathbf{U}} & \dot{\mathbf{V}} \\ \mathbf{P} & \mathbf{Q} \end{pmatrix} = \begin{pmatrix} \dfrac{1}{8\pi\mu}(\mathbf{1}\Delta - \boldsymbol{V}\otimes\boldsymbol{V})R\delta(t) & \boldsymbol{V}R\left(\dfrac{\delta(t)}{4\pi R^2}\right) \\[2ex] -\dfrac{kH(t)}{2\pi^{3/2}R^3}[(\mathrm{ex}\,z - \mathrm{Erf}(z))\mathbf{1} & -\dfrac{2H(t)C_v}{\pi^{3/2}R^4}\,\mathrm{ex}\,z^5 \\ \quad - \boldsymbol{V}R\otimes\boldsymbol{V}R(2\,\mathrm{ex}\,z^3 + 3\,\mathrm{ex}\,z - 3\,\mathrm{Erf}(z))] & \\[2ex] \dfrac{H(t)\boldsymbol{V}R}{2\pi^{3/2}R^2}(\mathrm{Erf}(z) - \mathrm{ex}\,z) & \dfrac{C_v H(t)}{k\pi^{3/2}R^3}\,\mathrm{ex}\,z^3 \end{pmatrix} \tag{67}
$$

where

$$
\mathrm{ex} = e^{-R^2/4C_v t}, \tag{68a}
$$

$$
z = \frac{R}{\sqrt{4C_v t}} \tag{68b}
$$

and

$$
\mathrm{Erf}(z) = \int_0^z e^{-t^2}\,dt. \tag{69}
$$

2. It is not necessary to keep Δt constant.

3. The Fourier transform formulation given in 3.3 is useful for investigating the behavior of various potentials. For example it is not difficult to obtain the boundary values of elastostatic simple and double layer potentials and their derivatives even for the most general case of anisotropy. See Nishimura and Kobayashi [14].

4. Biot's theory of consolidation is closely related to coupled thermoelasticity in that these theories are based on similar equations. For BIEM in thermoelasticity, the reader is referred to the paper by Sladek and Sladek [15]. This paper includes essentially the same BIE formulation as ours.

5. This chapter has presented BIEMs which use the time domain formulation. One may alternatively use some integral transformations on time and obtain integral representations for transformed quantities. An attempt along this line is found in a paper by Cheng and Liggett [16]. It appears to the present author, however, that the use of $\boldsymbol{u}$ and p in their governing equations makes their analysis simpler.

Acknowledgement

The author wishes to express his gratitude to Prof. T. Tamura for his criticism and to Mr. A. Umeda for his help in numerical work.

References

1 Terzaghi, K., Erdbaumechanik. F. Denticke, 1925

2 Biot, M.A., General theory of three-dimensional consolidation. J. Appl. Phys. 12, 155–164, 1941

3 Sandhu, R.S. and Wilson, E.L., Finite-element analysis of seepage in elastic media. Proc. ASCE, 95, EM, 641–652, 1969

4 Nishimura, N., A BIE formulation for consolidation problems. Boundary elements VII. C.A. Brebbia and G. Maier (eds.), 10, 47–55, Springer, 1985

5 Dubois, M. and Lachat, J.C., The integral formulation of boundary value problem. Variational Methods in Engineering. C.A. Brebbia and H. Tottenham (eds.), II, 9, 89–108, Southampton Univ. Press, 1973

6 Kobayashi, S. and Nishimura, N., On the indeterminancy of BIE solutions for the exterior problems of time-harmonic elastodynamics and incompressible elastostatics, Boundary Element Methods in Engineering. C.A. Brebbia (ed.), 282–296, Springer, 1982

7 Banerjee, P.K. and Butterfield, R., Boundary Element Methods in Engineering Science, McGraw Hill, 1981

8 Tamura, T., Eigenvalue problem of consolidation. Mem. Fac. Eng. Kyoto Univ., 42, 35–52, 1981

9 Predeleanu, M., Boundary integral method for porous media. Boundary Element Methods. C.A. Brebbia (ed.), 325–334, Springer, 1981

10 Kuroki, T., Ito, T. and Onishi, K., Boundary element method in Biot's linear consolidation. Appl. Math. Model., 6, 105–110, 1982

11 Garcia-Suarez, C. and Alarcon, E, Consolidation problems. Boundary Element Methods in Engineering. C.A. Brebbia (ed.), 377–390, Springer, 1982

12 Aramaki, G., Kuroki, T. and Onishi, K., Consolidation analysis by boundary element method. Boundary Element Methods in Engineering. C.A. Brebbia (ed.), 363–376, Springer, 1982

13 Aramaki, G., A novel boundary element method for thin layers in consolidation. Boundary Elements VII, C.A. Brebbia and G. Maier (ed.), 10, 3–12, Springer, 1985

14 Nishimura, N. and Kobayashi, S., Elastoplastic analysis by indirect methods. Developments in Boundary Element Methods–3. P.K. Banerjee and S. Mukherjee (eds.), Chapt. 3, Elsevier Appl. Sci. Publ., 1984

15 Sladek, V. and Sladek, J., Boundary integral equation method in thermoelasticity, part I: general analysis. Appl. Math. Model., 7, 241–253, 1983

16 Cheng, A.H-D. and Liggett, J.A., Boundary integral equation method for linear porous-elasticity with applications to soil consolidation. Int. J. Num. Meth. Eng., 20, 255–278, 1984

Chapter 4

A Review of Boundary Element Models of Saltwater Intrusion

by M. Kemblowski

Abstract

The Boundary Element Method (BEM) has been used for solving saltwater-intrusion problems for about six years. This paper attempts to review the use of the BEM for various hydrologic situations, which are described by different mathematical models: 1) steady-state cross sectional flow of freshwater above a stagnant saltwater zone; 2) transient cross sectional flow with the horizontally moving interface; 3) transient cross sectional flow with the vertically moving interface; and 4) transient horizontal flow, simplified by the use of the Dupuit-Forchheimer and Ghyben-Herzberg approximations. A brief discussion of each model's assumptions is also included. BEM models are concluded to cover a broad spectrum of saltwater-intrusion problems and to be superior to the finite-difference method (FDM) and the finite-element method (FEM) models for the moving boundary (interface) problems. The BEM seems also to be more efficient for analyzing horizontal-flow models, as long as the Ghyben-Herzberg approximation can be used. However, the FDM and the FEM are still better tools for analyzing the regional, transient saltwater-intrusion problems, for which this approximation cannot be used.

4.1 Introduction

Saltwater-intrusion phenomenon in coastal aquifers and areas underlain by saltwater-bearing formations is of practical importance. Analysis of the motion of the interface (or transition zone) between freshwater and saltwater should be performed while designing ground-water-recovery facilities in such areas. Perhaps the most general way to study this problem is to consider the variable-density hydrodynamic-dispersion model; however, the difficulties associated with the evaluation of the dispersive parameters and the numerical solution of the governing equations (mainly due to the numerical dispersion) make this model impractical in many cases.

One way to avoid these problems is to assume that the transition zone's thickness is small compared to the aquifer thickness and to introduce the so-called sharp-interface approximation. This approximation is used in all the Boundary Element Method (BEM) models described in this paper. Two types of models may utilize this approximation: 1) a three-dimensional flow model and 2) a horizontal flow model, which neglects the vertical-flow component. In the first part of this paper, the first type of model will be reviewed. The description of this type will be given, for simplicity, for the cross sectional plane. Three problems will be described in this part: 1) steady-state flow in the freshwater zone underlain by an immobile saltwater zone; 2) transient flow with horizontally moving interface; and 3) transient flow with vertically moving interface. In the second part a horizontal flow model will be reviewed. This model was formulated by utilizing the Dupuit-Forchheimer approximation, which assumes the flow in both zones to be basically horizontal. The Ghyben-Herzberg approximation, which assumes constant potential over the whole saltwater zone and therefore neglects flow in this zone, is also used in this model. The impact of these assumptions is analyzed.

4.2 Cross Sectional Flow Models

In this section we are considering three saltwater-intrusion models which do not use the Dupuit-Forchheimer approximation (Bear 1979). The first is a BEM solution of steady-state sea-water intrusion which utilizes the conformal transformation to the complex potential place $\phi - \psi$ (Liu and Liggett 1978). The second model (Liu et al. 1981) uses the BEM to solve the problem of horizontal interface motion in coastal aquifers and applies the explicit finite difference scheme to approximate the interface motion equation. The third solution combines the BEM and the implicitfinite difference scheme to solve the problem of vertical saltwater upconing (Kemblowski, 1984a, b).

It is worthwhile mentioning that although the three models are called cross sectional, the latter two may be easily extended to solve three-dimensional problems. The vertical upconing model may be also easily formulated in the radial system of coordinates $r - z$ and used to solve the problem of upconing under wells.

4.2.1 Mathematical Statement of the Problem

We are considering flow of two fluids in a homogeneous aquifer, assuming that a sharp interface separates them, such that each fluid occupies a separate zone (Fig. 1). The shapes of these zones may vary with time because of the interface motion. Assuming that the fluids and porous medium are incompressible, we can write for each zone a continuity equation (Bear 1979):

$$\nabla^2 \phi^F = 0 \tag{1}$$

$$\nabla^2 \phi^S = 0 \tag{2}$$

where

$$\phi^F = z + p/\gamma_f, \text{ freshwater potential} \tag{3a}$$

$$\phi^S = z + p/\gamma_S, \text{ saltwater potential} \tag{3b}$$

z = elevation above datum level

p = pressure

γ_F, γ_S = specific weights of freshwater and saltwater.

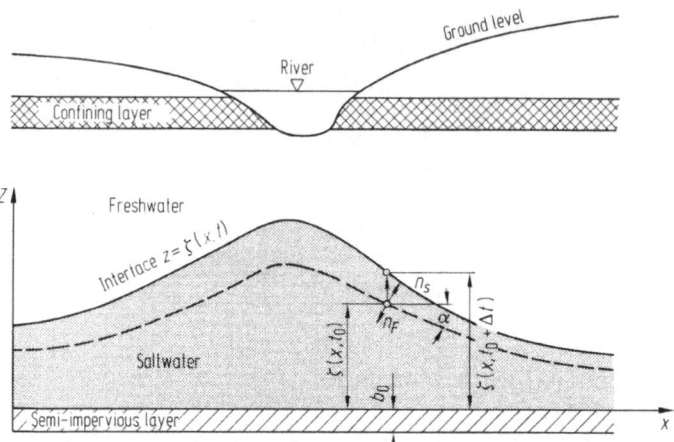

Fig. 1. Saltwater-freshwater interface upconing scheme

In order to be able to solve this system of partial differential equations [Eqs. (1) and (2)], boundary and initial conditions for each zone have to be specified. The initial conditions are given by initial location of the interface. The boundary conditions on the boundaries excluding the interface are of the same kind as for the flow of a single fluid and may be written as follows (Bear 1979)

$$\phi(\Gamma_1) = \phi_b \quad \text{(Dirichlet type)} \tag{4}$$

$$\partial\phi/\partial n(\Gamma_2) = -q \quad \text{(Neumann type)} \tag{5}$$

or

$$\frac{\partial\phi}{\partial n}(\Gamma_3) + a\phi(\Gamma_3) = c \quad \text{(mixed, Robbin type)} \tag{6}$$

where a and c are given functions, and Γ_1, Γ_2, Γ_3 are portions of the boundary Γ. The kinetic free-surface boundary conditions may also be specified on the free surface (Liggett and Liu 1983; Kemblowski 1984a).

The interface boundary conditions require the continuity of the pressure and the continuity of the fluxes normal to the interface and may be written as (Bear 1979)

$$\phi^S = \frac{\gamma_F}{\gamma_S}\phi^F + \frac{\gamma_S - \gamma_F}{\gamma_S}\zeta \tag{7}$$

$$\frac{\partial\phi^S}{\partial n_S} = -\frac{K_F}{K_S}\frac{\partial\phi^F}{\partial n_F} \tag{8}$$

where K_F and K_S are hydraulic conductivities for freshwater and saltwater and $\zeta(x, t)$ is the elevation of the interface.

The solution of Eqs. (1) and (2), subject to the boundary conditions [Eqs. (4), (5), and (6)] and interface relationships [Eqs. (7) and (8)], gives, in the transient case, nonzero normal fluxes along the interface. These fluxes cause interface movement. The interface-motion equation may be written as (Bear 1979)

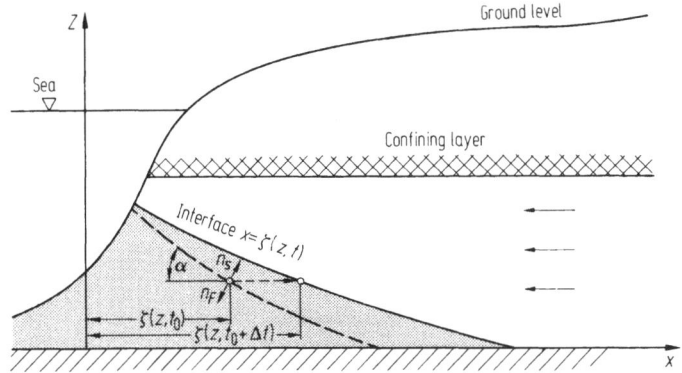

Fig. 2. Scheme of the horizontal motion of the interface

$$F(x, z, t) = 0. \tag{9}$$

The interface is a material surface (i.e., the fluid particles which, once on the interface, will remain on it unless they leave the system). Therefore, the material derivative of the interface equation is equal to zero (Bear 1979):

$$\frac{DF}{Dt} \equiv \frac{\partial F}{\partial t} + v \cdot \nabla F = 0 \tag{10}$$

where v is the particle velocity.

In general we may consider motion of the material suface in any direction. For practical purposes, we are interested, however, in the vertical ($x = $ const) or horizontal ($z = $ const) interface motion (Figs. 1 and 2). Appropriate equations may be written, respectively:

$$F = \zeta(x, t) - z = 0 \tag{11}$$

or

$$F = \xi(z, t) - x = 0. \tag{12}$$

The dot-product in Eq. (10) written for the horizontal motion of the interface may be expressed by

$$v \cdot \nabla F = |\nabla F| \text{ (projection of } v \text{ on } \nabla F) = |\nabla F| \, (\pm v_{n_F}) \tag{13}$$

where $v_{n_F} = $ particle velocity normal to the interface

$$v_{n_F} = -\frac{K_F}{S_o} \frac{\partial \phi^F}{\partial n_F} \tag{14}$$

where $S_o = $ effective porosity.

The positive and negative signs in Eq. (13) refer to declining or rising interface, respectively.

Substituting Eqs. (12) and (13) into Eq. (10) we obtain

$$\frac{\partial \xi}{\partial t} \pm \left(1 + \left(\frac{\partial \xi}{\partial z}\right)^2\right)^{1/2} \frac{K_F}{S_o} \frac{\partial \phi^F}{\partial n_F} = 0. \tag{15}$$

Substituting $\cot \alpha$ for $\dfrac{\partial \xi}{\partial z}$ we finally obtain the kinetic horizontal-motion equation of the interface:

$$\frac{\partial \xi}{\partial t} = \pm \frac{K_F}{S_o}(1 + \cot^2 \alpha)^{1/2} \frac{\partial \phi^F}{\partial n_F} = \pm \frac{K_F}{S_o \sin \alpha} \frac{\partial \phi^F}{\partial n_F}. \tag{16}$$

The equation for vertical motion of the interface may be derived in a similar manner. Its final form may be written as

$$\frac{\partial \zeta}{\partial t} = \frac{K_F}{S_o \cos \alpha} \frac{\partial \phi^F}{\partial n_F}. \tag{17}$$

Flow equations (1) and (2), along with the initial and boundary conditions, the interface relationships, and the interface motion equations, complete the mathematical statement of the problem.

4.2.2 Boundary Element Formulation

The Boundary Element Method is used to solve the governing equations [Eqs. (1) and (2)]. A detailed description of the method, as applied to ground-water flow, may be found in Liggett and Liu (1983). In the case of the two-dimensional Laplace equation, the boundary integral equation is given by

$$\alpha \phi(A) = \int_\Gamma \left[\phi(B) \frac{\partial}{\partial n}(\ln r) - \ln r \frac{\partial \phi(B)}{\partial n} \right] d\Gamma \tag{18}$$

where Γ is the boundary of the domain, B is a point on Γ, A is any point inside the domain or on its boundary, $\alpha = 2\pi$ if A lies inside the domain, $\alpha =$ interior angle of Γ if A is on Γ. The function $\ln r$ is free-space Green's function for Laplace's two-dimensional equation and r is the distance between A and B.

Using the boundary element approximation to calculate the boundary integral given by Eq. (18) in freshwater and saltwater zones leads to

$$\alpha_i \phi_i^F - \sum_{j=1}^{M} \left(\int_{\Gamma_F} \frac{1}{r_i} \frac{\partial r_i}{\partial n_F} N_j \, d\Gamma \right) \phi_j^F = - \sum_{j=1}^{M} \left(\int_{\Gamma_F} \ln r_i L_j \, d\Gamma \right) \left(\frac{\partial \phi^F}{\partial n} \right)_j ; \tag{19a}$$

$$i = 1, 2, \ldots, M$$

$$\alpha_i \phi_i^S = \sum_{j=M+1}^{N} \left(\int_{\Gamma_S} \frac{1}{r_i} \frac{\partial r_i}{\partial n_S} N_j \, d\Gamma \right) \phi_j^S = - \sum_{j=M+1}^{N} \left(\int_{\Gamma_S} \ln r_i L_j \, d\Gamma \right) \left(\frac{\partial \phi^S}{\partial n} \right)_j ; \tag{19b}$$

$$i = M + 1, \ldots, N$$

where

$N_j =$ basis function of node j for potential (Liu and Liggett 1983)
$L_j =$ basis function of node j for normal derivative of potential
$\phi_j =$ value of potential ϕ at node j
$\left(\dfrac{\partial \phi}{\partial n} \right)_j =$ value of normal derivative of potential ϕ at node j

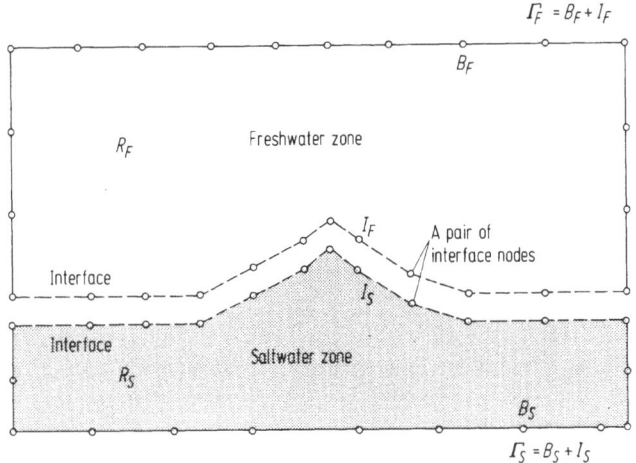

Fig. 3. Discretized region for the numerical solution

r_i = the distance between node i and an integration point
Γ_F = boundary of freshwater zone
Γ_S = boundary of saltwater zone
M = number of nodes in freshwater zone
N = total number of nodes.

Note that each node on the interface represents actually a pair of nodes and is numbered twice, once in the freshwater zone and once in the saltwater zone (Fig. 3). If the basis functions are assumed linear, then the integrations in Eqs. (19a) and (19b) are performed analytically.

The system of N equations given by Eqs. (19a) and (19b) has $2N$ unknowns (namely ϕ and $\partial\phi/\partial n$ at each node). However, if the number of interface node pairs is K, then in a well-posed problem, a value of either ϕ or $\partial\phi/\partial n$ or their relationship is given [Eqs. (4), (5), and (6)] at $2(N - K)$ noninterface nodes. These boundary conditions increase the number of equations to $2N - 2K$. Additional $2K$ equations are provided by the interface relationship [Eqs. (7) and (8)] for each pair of the interface nodes.

4.2.3 Steady-state Sea-Water Intrusion

The BEM solution described herein was presented by Liu and Liggett (1978). The BEM was applied to study the steady-state location of the interface in a confined aquifer. The physical situation under consideration is depicted in Fig. 4, which shows a freshwater canal running in the middle of a flat with the freshwater seeping from the canal and thus preventing the saltwater intrusion. Charmonman (1966) introduced the following dimensionless variables:

$$(x', z') = 2N/Q_c(x, z) \qquad (20a)$$

$$(\phi', \psi') = 2/Q_c(\phi, \psi) \qquad (20b)$$

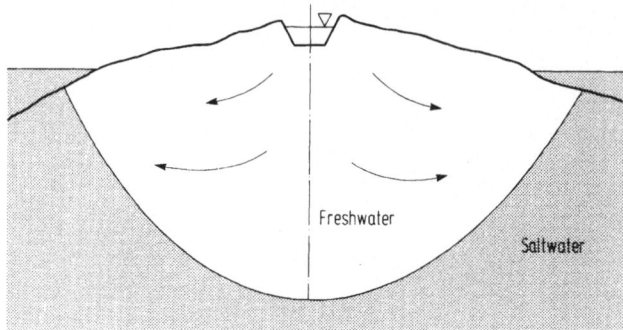

Fig. 4. Hydrologic scheme of the steady-state saltwater intrusion

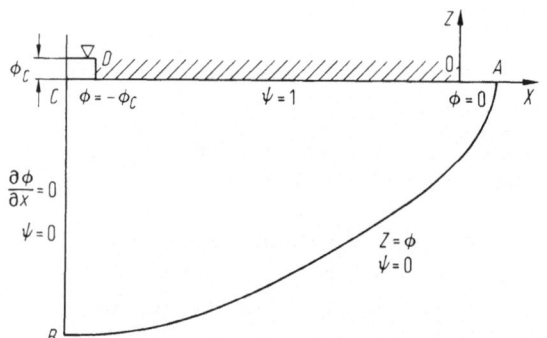

Fig. 5. Simplified scheme of the steady-state saltwater intrusion

where $N = K_F \dfrac{\gamma_S - \gamma_F}{\gamma_F}$, ψ is the stream function, and Q_c is the rate of seepage from the channel, $\phi = -\phi^F$. The primes will be dropped for convenience.

The potential ϕ and stream function ψ are related to the fluxes by

$$q_x = \frac{\partial \phi}{\partial x} = \frac{\partial \psi}{\partial z} \tag{21a}$$

$$q_z = \frac{\partial \phi}{\partial z} = -\frac{\partial \psi}{\partial x}. \tag{21b}$$

The problem may be now simplified to the model depicted in Fig. 5 and described by either of the following mass balance equations:

$$\nabla^2 \phi = 0 \tag{22}$$

$$\nabla^2 \psi = 0. \tag{23}$$

The boundary conditions of the problem may be written in terms of ϕ or ψ as follows (Fig. 5)

$$\phi = -\phi_c \qquad \frac{\partial \psi}{\partial n} = 0 \quad \text{on } C - D \tag{24a}$$

$$\frac{\partial \phi}{\partial n} = 0 \qquad \psi = 1 \quad \text{on } D - O \tag{24b}$$

$$\phi = 0 \qquad \frac{\partial \psi}{\partial n} = 0 \quad \text{on } O - A \tag{24c}$$

$$\phi = z \qquad \psi = 0 \quad \text{on } A - B \tag{24d}$$

$$\frac{\partial \phi}{\partial n} = 0 \qquad \psi = 0 \quad \text{on } B - C. \tag{24e}$$

Note that the location of the interface, and particularly of point B, is unknown, and therefore the problem cannot be solved directly for either ϕ or ψ. In order to at least partially avoid this problem, the complex potential $W = \phi + i\psi$ is introduced, whose property is

$$\frac{dW}{dz} = q_x - iq_y \tag{25}$$

where $Z = x + iz$.

The problem described by Eqs. (22) and (23) may now be represented in the complex potential plane $\phi - \psi$ (Fig. 6), with the dependent variables $x(\phi, \psi)$ and $z(\phi, \psi)$ as follows

$$\frac{\partial^2 x}{\partial \phi^2} + \frac{\partial^2 x}{\partial \psi^2} = 0 \tag{26}$$

$$\frac{\partial^2 z}{\partial \phi^2} + \frac{\partial^2 z}{\partial \psi^2} = 0. \tag{27}$$

The variables x and z are related by the Cauchy-Riemann conditions

$$\frac{\partial x}{\partial \phi} = \frac{\partial z}{\partial \psi}; \quad \frac{\partial x}{\partial \psi} = -\frac{\partial z}{\partial \phi}. \tag{28}$$

The saltwater-intrusion problem is solved directly for $z(\phi, \psi)$ and then the solution

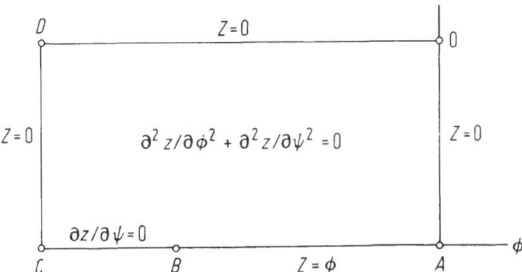

Fig. 6. The steady-state problem in the complex potential plane

for $x(\phi, \psi)$ is obtained by integrating (28). The boundary conditions for $z(\phi, \psi)$ are
(Fig. 6)

$$z = 0 \quad \text{for } \psi = 1 \tag{29a}$$

$$z = 0 \quad \text{for } \phi = 0 \tag{29b}$$

$$z = \phi \quad \text{for } \phi_B < \phi < 0 \tag{29c}$$

$$\frac{\partial z}{\partial n} = 0 \quad \text{for } \phi_C < \phi < \phi_B. \tag{29d}$$

For such a problem the BEM may be readily used. However, since the location of
point B is chosen arbitrarily, the resulting horizontal dimensions, obtained from
the inverse Cauchy-Riemann equations, may be quite different than those of the
considered problem. Therefore, if a particular problem with given geometry is
considered, the final solution is obtained by varying the value of ϕ_B until the
estimated horizontal dimensions are close enough to the real values.

4.2.4 Horizontal Motion of the Interface – Explicit Solution

The interface horizontal-motion equation may be numerically approximated using
the explicit finite-difference scheme and written as (Liu et al. 1981)

$$\zeta^{k+1} = \zeta^k \pm \frac{K_F \Delta t}{S_o \sin \alpha} \left(\frac{\partial \phi^F}{\partial n_F} \right)^k. \tag{30}$$

For this approximation of the motion equation the solution procedure for each
time step (Δt) consists of:

1) Solving Eqs. (1) and (2) [approximated by Eqs. (19a) and (19b)], subject to the
boundary and initial conditions [Eqs. (4), (5), and (6)] and interface relationships
[Eqs. (7) and (8)]. As a result, the values of the potentials $(\phi^F)^k$ and $(\phi^S)^k$ and their
normal derivatives $(\partial \phi^F / \partial n_F)^k$ and $(\partial \phi^S / \partial n_S)^k$ are obtained.
2) Calculating a new location of the interface nodes using equation 30. Usually the
boundary conditions and interface relationships are incorporated into the system
of BEM equations [Eqs. (19a) and (19b)] during its formulation, thus the actual
number of equations and unknowns is equal to N (the total number of nodes).

4.2.5 Vertical Motion of the Interface – Implicit Solution

The interface vertical-motion equation may be approximated in the same manner
as the horizontal-motion equation and written as

$$\zeta^{k+1} = \zeta^k + \frac{K_F \Delta t}{S_o \cos \alpha} \left(\frac{\partial \phi^F}{\partial n_F} \right)^k. \tag{31}$$

However, the use of this approximation may be limited to short time steps (Δt),
due to the oscillations and instability associated with the explicit finite difference
scheme for longer time steps (Kemblowski 1984b). For these reasons Kemblowski
(1984b) developed for the vertical motion of the interface an implicit solution, which
avoids the aforementioned problems.

Using the implicit finite difference scheme to approximate Eq. (17) leads to

$$\zeta^{k+1} = \zeta^k + \frac{K_F \Delta t}{S_o \cos \alpha} \left\{ \omega \left(\frac{\partial \phi^F}{\partial n_F} \right)^{k+1} + (1 - \omega) \left(\frac{\partial \phi^F}{\partial n_F} \right)^k \right\} \tag{32}$$

where ω = the weighting factor.

Equation (32) contains a term which is unknown for time $k\Delta t$, namely $\left(\frac{\partial \phi^F}{\partial n_F} \right)^{k+1}$.

This term has to be "predicted" in order to use the implicit scheme. The prediction procedure is derived in the following manner. After writing Eq. (7) for times $k\Delta t$ and $(k + 1)\Delta t$ the former is subtracted from the latter which gives

$$(\phi^F)^{k+1} = (\phi^F)^k + \frac{\gamma_S}{\gamma_F} \{ (\phi^S)^{k+1} - (\phi^S)^k \} - \frac{\gamma_S - \gamma_F}{\gamma_F} (\zeta^{k+1} - \zeta^k). \tag{33}$$

Substituting Eq. (32) into Eq. (33) leads to

$$(\phi^F)^{k+1} = (\phi^F)^k - \frac{\gamma_S}{\gamma_F} \{ (\phi^S)^{k+1} - (\phi^S)^k \}$$

$$- \frac{\gamma_S - \gamma_F}{\gamma_F} \frac{K_F \Delta t}{S_o \cos \alpha} \left\{ \omega \left(\frac{\partial \phi^F}{\partial n_F} \right)^{k+1} + (1 - \omega) \left(\frac{\partial \phi^F}{\partial n_F} \right)^k \right\}. \tag{34}$$

This equation and the flux-continuity condition

$$\left(\frac{\partial \phi^F}{\partial n_F} \right)^{k+1} = -\frac{K_S}{K_F} \left(\frac{\partial \phi^S}{\partial n_S} \right)^{k+1} \tag{35}$$

provide the interface relationships necessary for prediction of $\left(\frac{\partial \phi}{\partial n} \right)^{k+1}$. Notice that all the terms for time $k\Delta t$ are known in these equations. The solution procedure for this implicit scheme consists of the following:

1) Solving Eqs. (1) and (2) [approximated by Eqs. (19a) and (19b)] subject to the boundary and initial conditions [Eqs. (4), (5), and (6)] and interface relationships [Eqs. (7) and (8)], to obtain

$$(\phi^F)^k, (\phi^S)^k, \left(\frac{\partial \phi^F}{\partial n_F} \right)^k, \left(\frac{\partial \phi^S}{\partial n_S} \right)^k.$$

2) Solving Eqs. (1) and (2) subject to the boundary and initial conditions and the interface relationships given by Eqs. (34) and (35) to obtain the predicted values of $(\phi^F)^{k+1}, (\phi^S)^{k+1}, (\partial \phi^F/\partial n_F)^{k+1}, (\partial \phi^S/\partial n_S)^{k+1}$.

3) Calculating a new location of the interface nodes using Eq. (32).

4.3 Horizontal Flow Models

The principal difference between the so-called cross sectional models (which, in fact, may be considered three dimensional) and the horizontal-flow models is that the latter assume that the ground-water flow is basically horizontal (the Dupuit-Forchheimer approximation). The validity of this assumption has been generally accepted, except for problems which involve ground-water flow in the vicinity of partially penetrating sinks or sources, particularly if the Dirichlet-type boundary condition is specified at these singularities. A further simplifying approximation used for saltwater-intrusion problem is that no flow occurs in the saltwater zone and its potential is constant. In other words, it assumes that the response time of the flow in the saltwater zone and interface motion is of the same order as the fluctuations of the potential in the freshwater zone. Thus, at any given time, the interface location satisfies the following relationship:

$$\zeta = \frac{\gamma_S}{\gamma_S - \gamma_F}\phi_o^S - \frac{\gamma_F}{\gamma_S - \gamma_F}\phi^F \tag{36}$$

where ϕ_o^S is the saltwater potential, constant in the saltwater zone.

4.3.1 Mathematical Statement of the Problem

Consider flow in a homogeneous confined aquifer as shown in Fig. 7. A solution domain is divided into two parts: Region 1 which consists of freshwater underlain by the bedrock, and Region 2, consisting of the freshwater underlain by the immobile saltwater. Assuming the Dupuit-Forchheimer and Ghyben-Herzberg approximation valid, the mass-balance equations may be written for the two regions as follows:

$$K_F \nabla (b \nabla \phi_1^F) = S_i \frac{\partial \phi_1^F}{\partial t} + Q \tag{37}$$

$$K_F \nabla \left\{ \left(b - \frac{\gamma_S}{\Delta \gamma}\phi_o^S + \frac{\gamma_F}{\Delta \gamma}\phi_2^F \right) \nabla \phi_2^F \right\} = S_o \frac{\gamma_F}{\Delta \gamma} \frac{\partial \phi_2^F}{\partial t} + Q \tag{38}$$

where Q is the unit recharge or withdrawal, b is the aquifer thickness, and S is the aquifer storativity.

For steady-state flow problems these equations may be combined into one, which describes flow in the two regions:

$$K_F \nabla^2 \phi = Q \tag{39}$$

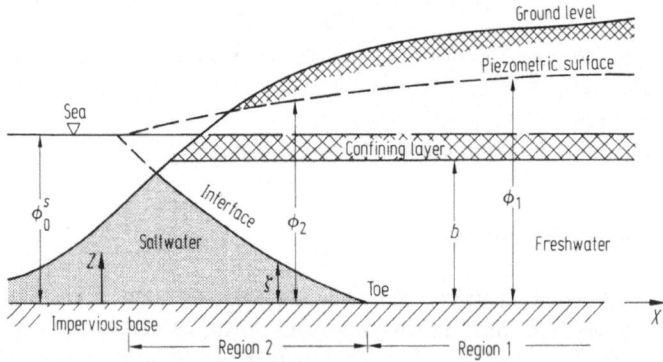

Fig. 7. Saltwater intrusion into a confined aquifer-horizontal flow model (after Taigbenu et al. 1981)

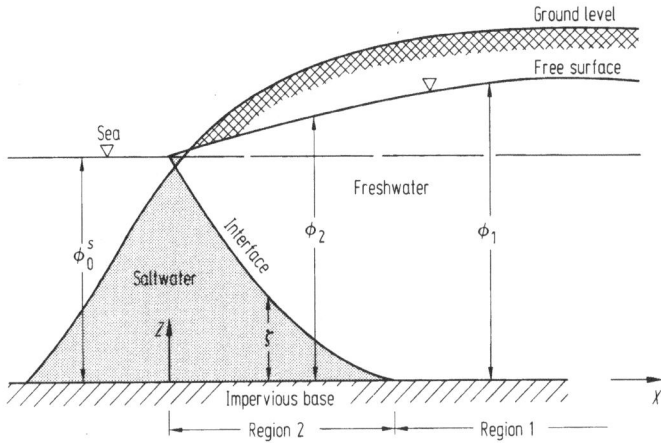

Fig. 8. Saltwater intrusion into a unconfined aquifer-horizontal flow model (after Taigbenu et al. 1981)

where the special potential ϕ is defined by

$$\phi = b\left(\phi_1^F + \frac{\Delta\gamma}{2\gamma_F}b - \frac{\gamma_S}{\gamma_F}\phi_o^S\right) \qquad \text{in Region 1} \qquad (40)$$

$$\phi = \frac{\gamma_F}{2\Delta\gamma}\left(\phi_2^F + \frac{\Delta\gamma}{\gamma_F}b - \frac{\gamma_S}{\gamma_F}\phi_o^S\right)^2 \qquad \text{in Region 2.} \qquad (41)$$

At the boundary of Region 1 and Region 2 ($\zeta = 0$), as well as elsewhere, the potential ϕ and its derivatives are continuous. Notice that this special potential cannot be used to combine the transient flow equations in the two regions for the unconfined case. However, the potential ϕ may be used to simplify the transient flow equation, if the area under consideration is of the Region 2 type (i.e., when saltwater zone is present the freshwater zone in the entire area). This simplified equation may be written as follows

$$K_F\nabla^2\phi = S_o\left(\frac{2\gamma_F}{\Delta\gamma}\right)^{1/2}\frac{\partial\phi^{1/2}}{\partial t} + Q. \qquad (43)$$

In this form, the equation cannot be solved by the BEM because of the nonlinear form of the time derivative. The linearization may be performed in two ways:

1)
$$\phi_1^* = (2\phi)^{1/2} \qquad (44)$$

$$K_F\bar{\phi}_1^*\nabla^2\phi_1^* = S_o\left(\frac{\gamma_F}{2\Delta\gamma}\right)\frac{\partial\phi_1^*}{\partial t} + Q \qquad (45)$$

2)
$$\partial\phi^{1/2}/\partial t = 1/(2\phi^{1/2})\partial\phi/\partial t \qquad (46)$$

$$K_F\nabla^2\phi = S_o\left(\frac{\gamma_F}{2\Delta\gamma\bar{\phi}}\right)^{1/2}\frac{\partial\phi}{\partial t} \qquad (47)$$

where $\bar{\phi}^*, \bar{\phi}$ are regional averages of the special potentials ϕ^* and ϕ.

For the unconfined flow the transient mass-balance equation in the two regions may be written as follows

$$K_F \nabla(\phi_1^F \nabla \phi_1^F) = S_o \frac{\partial \phi_1^F}{\partial t} + Q \tag{48}$$

$$K_F \nabla \left\{ \left(\phi_1^F - \frac{\gamma_S}{\Delta\gamma} \phi_o^S + \frac{\gamma_F}{\Delta\gamma} \phi_2^F \right) \right\} \nabla \phi_2^F = S_o \frac{\gamma_S}{\Delta\gamma} \frac{\partial \phi_2^F}{\partial t} + Q. \tag{49}$$

In this case the special potential ϕ is given by

$$\phi = \frac{1}{2} \left((\phi_1^F)^2 - \frac{\gamma_S}{\gamma_F}(\phi_o^S)^2 \right) \quad \text{in Region 1} \tag{50}$$

$$\phi = \frac{1}{2} \frac{\gamma_S}{\Delta\gamma}(\phi_2^F - \phi_o^S)^2 \quad \text{in Region 2.} \tag{51}$$

This special potential may again be used to either combine Eqs. (48) and (49) for the steady-state conditions:

$$K_F \nabla^2 \phi = Q \tag{52}$$

or to simplify the transient flow equation in the Region-2-type aquifer

$$K \nabla^2 \phi - Q = S_o \left(\frac{2\gamma_S}{\Delta\gamma} \right)^{1/2} \frac{\partial \phi^{1/2}}{\partial t}. \tag{53}$$

Either of the linearization procedures may now be used to linearize Eq. (53).

4.3.2 Boundary Element Formulation

The steady-state and linearized transient equations may be written in a non-dimensional form as (Taigbenu et al. 1984)

$$\nabla^2 \phi - Q/K = 0 \tag{54}$$

for the steady state case, and

$$\nabla^2 \phi - Q/K = \frac{\partial \phi}{\partial t} \tag{55}$$

for the transient case.

Applying the BEM to Eq. (54) leads to

$$\alpha\phi(A) = \int_\Gamma \left[\phi(B) \frac{\partial}{\partial n}(\ln r) - \ln r \frac{\partial \phi(B)}{\partial n} \right] d\Gamma + \int_\Omega \frac{Q}{K} \ln r \, d\Omega. \tag{56}$$

The only difference between Eq. (56) and Eq. (18) is the areal integral on the right-hand side. Thus, the boundary integration is carried on in the same manner as described previously. The areal integration may be performed numerically by dividing the domain Ω into a number of subdomains (for instance triangular elements) and approximating the distribution of the integrand $Q \ln r/K$ with the linear basis function, in a way similar to the finite-element approximation. In many cases $Q(x, y)$ is nonzero only at a number of points (wells), and the areal integration is not necessary.

In the case of the parabolic equation [Eq. (55)], the Laplace transform may be used to remove the time derivative. The Laplace transform is defined as

$$\tilde{f}(x, y, s) = \int_0^\infty f(x, y, t)e^{-st}\, dt \tag{57}$$

where s is the transform parameter. Application of the Laplace transform to the potential ϕ and recharge Q in the parabolic equation leads to

$$\Delta^2\tilde{\phi} - s\tilde{\phi} = \tilde{Q}/K - \phi_o \tag{58}$$

where ϕ_o is the initial distribution of ϕ. The boundary-element representation of this equation is given by

$$\alpha_i\tilde{\phi}_i - \sum_{j=1}^N \left(\int_\Gamma s^{1/2}N_jK_1(s^{1/2}r)\frac{\partial r_i}{\partial n}\, d\Gamma \right)\tilde{\phi}_j$$

$$= \sum_{j=1}^N \left(\int_\Gamma K_0(s^{1/2}r)L_j\, d\Gamma \right)\left(\frac{\partial\tilde{\phi}}{\partial n}\right)_j + \int_\Gamma \left(\frac{Q}{K} - \phi_o\right)\ln r\, d\Omega;$$

$$i = 1, 2, \ldots, N \tag{59}$$

where $K_0(\alpha)$ is the modified Bessel function of the second kind and order zero, and $K_1(\alpha) = -dK_0/d\alpha$, the modified Bessel function of the second kind and order one.

The integrations in Eq. (59) have to be performed numerically. The solution procedure consists of:

1) Solving Eq. (59) for $\tilde{\phi}$ and $\partial\tilde{\phi}/\partial n$ at the boundary nodes for six to twelve (L) selected values of the transform parameter s.
2) Calculation of $\tilde{\phi}(x, y, s_k)$ at the interior points of interest, using Eq. (59). For this use, subscript i in Eq. 59 refers to the points of interest.
3) Inverting the obtained results to the time domain. The inversion uses the assumption of exponential behavior of the potential ϕ in time, which may be written as (Liggett and Liu 1983)

$$\phi(x, y, t) = \phi_S(x, y) + \sum_{l=1}^L a_le^{-s_kt}; \quad k = 1, 2, \ldots, L \tag{60}$$

where ϕ_S is the steady-state solution of Eq. (55), or the transformed solution for $s = 0$, and vector a_l is to be determined from the Laplace transform of Eq. (60):

$$\tilde{\phi}(x, y, s_k) = \frac{\tilde{\phi}_S}{s_k} + \sum_{l=1}^L \frac{a_l}{s_l + s_k}; \quad k = 1, 2, \ldots, L \tag{61}$$

which gives a system of L linear equations with L unknowns (s_k). After solving Eq. (61) at any given point for a_l, the potential ϕ may be calculated at any time using Eq. (60).

4.4 Discussion and Conclusions

The models reviewed in this paper cover a broad spectrum of saltwater-intrusion problems. In this section some advantages and disadvantages of them will be discussed. With regard to the first type of models (cross sectional), the BEM has a number of advantages compared to either the Finite Difference Method (FDM) or to the Finite Element Method (FEM). First, by substituting the boundary integral (either line or surface integral) for the areal or volumetric integral, the BEM reduces the dimensions of the problem by one. Secondly, although FEM or FDM solutions to the moving surface problem are possible, the bookkeeping of the time-varying numerical network and the computer storage requirements make these solutions often impractical. In the case of the FDM, approximating the interface boundary with the rectangular network is difficult. Also the FEM solution does not directly provide the distribution of the normal derivative of the potential ($\partial \phi / \partial n$) along the boundaries. This distribution is obtained by solving an additional system of equations. All these disadvantages make the FDM and the FEM inferior compared to the BEM for solving the moving boundary problem. Let us now consider each model of the first type. The first, steady-state cross sectional flow model was formulated in order to calculate the distance from the freshwater canal to the sea and the canal's width for a given freshwater potential at the canal bottom (ϕ_c) and the location of the interface under the center line of the canal (z_B). For such a problem the first model is very well suited. However, if the geometry and ϕ_c are given and one wants to calculate the location of the interface, then the vertical-motion model is more suitable and it may be used to obtain either transient or steady-state results.

The second model, which simulates the horizontal motion of the interface, is the only one from this family that can simulate the motion of the toe (i.e., intersection of the interface and the bedrock). The explicit scheme used for the approximation of the interface motion equation may cause instability of the numerical solution for longer time steps, therefore developing an implicit solution for the moving interface is desirable. Such solution technique could also be used in other areas, for example in petroleum and chemical engineering.

The third model, which deals with the vertical motion of the interface, may be used for the analysis of the upconing phenomenon, whether it is caused by drains or rivers or wells. The implicit scheme developed for this model assures the stability of the solution.

The latter two of the so-called cross sectional models may be easily extended to simulate three-dimensional problems (Liggett and Liu 1983). However, if the horizontal dimensions of the aquifer are much bigger than its thickness, then the number of nodes may be prohibitively large; besides usually under such circumstances the Dupuit-Forchheimer approximation may be introduced. Thus, these models should be mainly used for the cross sectional flow, especially in the vicinity of partially penetrating sinks or sources.

The horizontal-flow models reviewed in this paper are particularly well suited for the analysis of the steady-state flow. The Dupuit-Forchheimer approximation used in these models does not cause substantial errors as long as there are no sinks or sources with the Dirichlet-type boundary condition in the region. If such singularities exist, the estimation of the interface location may be erroneous due to the pronounced vertical component of the flow and, associated with it, the difference of the potential between the sink-source and the interface.

With regard to the transient flow, there are several problems related to the use of the BEM horizontal-flow models. Firstly, they cannot simulate the toe movement, which is usually of interest in the regional flow modeling. Secondly, the Ghyben-Herzberg approximation, which is used in the model, may for some cases lead to erroneous results, especially when the saltwater flow does not respond immediately to the freshwater potential fluctuations, as happened in the case of saltwater upconing under a river analyzed by Kemblowski (1984a). In this case the source of the saltwater intrusion was a saltwater aquifer separated from the overlying alluvial aquifer by a semi-confining layer. The immediate response of the saltwater flow and potential to the freshwater-potential fluctuations was prevented by the semi-confining layer, and the simulation results clearly indicate that the Ghyben-Herzberg approximation is not acceptable for the analysis of this system's behavior. In addition the areal integration involved in the BEM solution lessens the advantage of this solution over the finite-difference or the finite-element methods. Therefore, for the analysis of the regional transient problems of saltwater intrusion either the finite-difference method or the finite-element method is still a better approximation, because neither of them has to utilize the Ghyben-Herzberg approximation.

Acknowledgements

This work was partially supported by the U.S. Department of the Interior, Bureau of Reclamation, under grant 4-FG-93-00080.

References

Bear, J. (1979). Hydraulics of Groundwater. McGraw-Hill, Inc. New York.

Liu, P.L.-F., and J.A. Liggett (1978). An Efficient Numerical Method of Two-Dimensional Steady Groundwater Problems. Water Resources Research, 14(3): 385–390.

Liu, P.L.-F., A.H.-D. Cheng, J.A. Liggett, and J.H. Lee (1981). Boundary and Integral Equation Solutions to Moving Interface Between Two Fluids in Porous Media. Water Resources Research, 17(5): 1445–1452.

Kemblowski, M. (1984a). Saltwater Upconing Under a River – A Boundary-Element Solution. Proceedings of the 6th International Conference on Boundary Element Methods in Engineering, QE2.

Kemblowski, M. (1984b). Saltwater-Freshwater Transient Upconing – An Implicit Boundary-Element Solution. Accepted for publication in Journal of Hydrology.

Liggett, J.A., and O.L.-F. Liu (1983). The Boundary Integral Equation Method for Porous Media Flow. Allen and Unwin, London.

Charmonman, S. (1966). A Numerical Method of Solution of Free Surface Problems. Journal of Geophysical Research, 71(16): 3861–3868.

Taigbenu, A.E., J.A. Liggett, and A.D.-D. Cheng (1984). Boundary Integral Solution fo Seawater Intrusion Into Coastal Aquifers. Water Resources Research, 20(8): 1150–1158.

Chapter 5

Boundary Element Modelling of Interface Phenomena

by A.P.S. Selvadurai and M.C. Au

Abstract

In this article we examine the application of the boundary element method to the study of the non-linear interface behaviour between two material regions. The non-linear interface response is modelled either by Coulomb frictional behaviour or by interface plasticity. An incremental formulation is adopted for the analysis of the non-linear pheonomena. The incremental non-linear analysis is used to examine the two-dimensional problem of a finite elastic region which contains a circular rigid inclusion. The numerical results presented in the paper illustrates the manner in which the non-linear phenomena at the inclusion-elastic medium interface contributes to the global non-linear responses in the composite.

5.1 Introduction

The category of problems which examine the mechanical behaviour of contact regions form an important branch of applied mechanics which has extensive engineering applications. The results of such research has applications in the study of mechanics of composite materials, tribology, soil-structure interaction, mechanical response of fractured interfaces and in the examination of damage modelling in micro-mechanics. In modelling these problems attention is usually focussed on establishing the interface response on either global or local responses of the bodies in contact. These include the assessment of load-displacement responses or contact stiffnesses, stress concentrations and the identification of regions which experience separation, slip and frictional adhesion either in the contact regions or within the material bodies. The research in this area has attracted a considerable amount of concerted effort and accounts of current developments in this area are given by Duvaut and Lions [1], Gudehus [2], Desai and Christian [3], de Pater and Kalker, [4], Selvadurai [5], Gladwell [6], Kikuchi and Oden [7], Panagiotopoulos [8], Johnson [9], and Selvadurai and Voyadjis [10].

Material interfaces usually exhibit a multitude of interface phenomena which are influenced by the mechanical properties of the media in contact, the surface characteristics at the macro- and micro-scale and the rate at which loads are applied to and transferred between the interfaces (see e.g., Bowden and Tabor [11], Johnson [9]). In its simplest form, the response of interfaces under static loadings can be characterized by completely smooth or completely bonded behaviour with either elastic-plastic, Coulomb friction or finite friction occupying an intermediate position. Other types of interface responses can include dilatant phenomena, strain hardening and strain softening responses and time-dependent effects of a visco-elastic or viscoplastic nature. The analysis of interface contact problems can be approached via a variety of methods. These include analytical schemes and numerical schemes based on finite element, boundary element and boundary integral equation schemes. The analytical study of contact problems invariably focus on the treatment of simplified interface responses encountered at frictionless and bonded regions. Certain aspects of unilateral contact problems, at frictionless elastic interfaces, which involve advancing or receding contact regions can also be examined via the analytical method. Detailed expositions of these problems are given by Uflyiand [12], Galin [13], Dundurs and Stippes [14], de Pater and Kalker [4], Sneddon [15], Selvadurai [5] and Gladwell [6]. The analytical treatment of contact problems related to interfaces which exhibit Coulomb friction or finite friction is quite complicated. For example, the linear elastic problem concerning the frictional contact between a rigid circular punch and an elastic halfspace can be examined only via an incremental formulation of the governing integral equations (see e.g., Spence [16], de Pater and Kalker [4], Kalker [17], Turner [18] and Galdwell [6]). Such analytical schemes for the analysis of contact problems with non-linear interface phenomena can be successfully applied only in a limited number of situations which relate to problems with simplified geometrical and loading situations.

The complexities of the analytical methods of analyses have prompted the application of numerical methods of stress analyses, such as the finite element techniques, to the study of contact problems with interface non-linearities. Special interface elements have been extensively employed in the finite element analysis of contact problems in structural mechanics and geomechanics. Among the earliest investigations in this area are due to Goodman et. al. [19], Goodman and Dubois [20], Ghabbousi et. al. [21], Zienkiewicz [22], and Fredriksson [23]. Recent pertinent articles include those by Boulon et. al. [24], Pande and Sharma [25], Ito et. al. [26] and Selvadurai and Faruque [27]. In recent studies Desai and co-workers [28, 29] have employed interface elements for the study of both static and dynamic soil-structure interaction. These studies also take into consideration softening and hardening processes encountered during cyclic loading of geological interfaces. Detailed surveys of the application of interface elements to finite element analysis of interface and contact problems are given by Goodman [30], Desai [31], Herrmann [32], Wilson [33], Gaertner [34], Okamoto and Nabazawa [35] and Zienkiewicz [36].

The boundary element method offers a further numerical scheme which can be used quite effectively for the study of non-linear interface phenomena. Andersson [37, 38] and Andersson and Allan-Persson [39] have applied the boundary element method to examine the interface frictional contact and separation in two-dimensional problems involving bearings and other mechanical assemblages. Paris and Garrido [40] has investigated the class of frictional contact problems involving two-dimensional plane punches and half-plane regions. Selvadurai and Au [41] have examined the problem of interface separation, interface frictional contact and interface slip at the boundary of rigid spherical inclusions embedded in elastic media. References to further studies involving boundary element treatment of non-linear material and interface phenomena are given by Brebbia [42], Telles [43], Banerjee and Butterfield [44], and Brebbia et. al. [45]. In this article we examine the application of the boundary element method to the study of certain two-dimensional problems involving an inclusion reinforced elastic medium. It is assumed that non-linear phenomena occur at the inclusion-elastic medium interface. The interface non-linearities include either Coulomb-type frictional phenomena or plasticity phenomena. The inclusion-reinforced elastic medium is subjected to tractions or displacements at the external boundary. The resulting boundary defor-mations are interpreted as the deformational responses of the inclusion-reinforced body. An iterative numerical scheme is used to examine the frictional contact problem. The numerical results presented in the article indicate the manner in which non-linear responses at the inclusion-elastic medium interface can contribute to the development of bulk non-linear irreversible phenomena in the composite.

5.2 Basic Equations

We focus attention on the class of plane problems involving an isotropic elastic medium which is bounded internally by a rigid inclusion (Fig. 1). The outer boundary of the elastic medium is subjected to specific boundary deformations. Such deformations can include either constant or uniform boundary displacements which effectively model homogeneous states. The general form of the rigid body displacement experienced by the inclusion region Ω_b can be expressed in the form of the vector $\{v_x, v_y, \omega\}$ where v_x and v_y are the components of the displacement vector in the x- and y-directions respectively and ω is the rigid body rotation of Ω_b about a point which is $-\{r_0\}$ from its boundary $\partial\Omega_b$. The isotropic elastic medium (Ω_m) surrounding the rigid inclusion satisfies the equilibrium equations

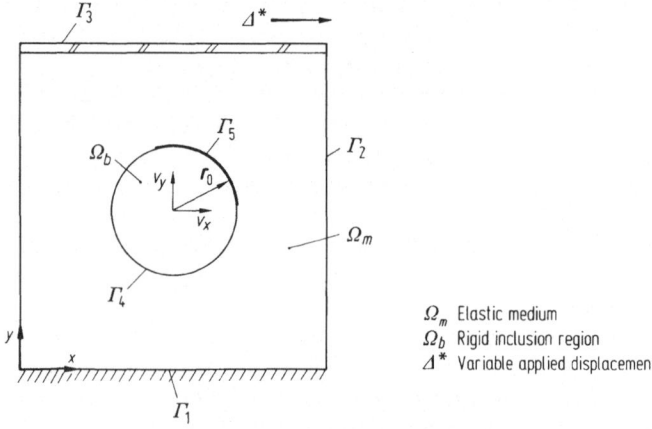

Ω_m Elastic medium
Ω_b Rigid inclusion region
Δ^* Variable applied displacement

Fig. 1. A unit cell with a rigid inclusion

$$\sigma_{ij,j} = 0; \quad x_i \in \Omega_m \tag{1}$$

where σ_{ij} $(i, j = 1, 2)$ is the two-dimensional state of stress in the elastic medium. The boundary of the elastic medium $\partial\Omega_m$ can be its subsets $\Gamma_1, \Gamma_2, \ldots, \Gamma_5$ (Fig. 1). The boundary of the inclusion region $\partial\Omega_b = \Gamma_4 \cup \Gamma_5$. It is assumed that the inclusion-elastic medium boundary exhibits non-linear interface phenomena. The outer boundary of the composite region is subjected to specified tractions or displacements. During the application of these tractions or displacements non-linear processes occur at the interface $\partial\Omega_b$. Processes such as frictional locking contact and slip are assumed to occur without the occurrence of separation. This constraint can be satisfied by imposing a priori a suitable isotropic confining stress. Considering Fig. 1 we have

$$u_i = \bar{u}_0; \quad x_i \in \Gamma_1 \tag{2}$$

$$t_i = \bar{t}_0; \quad x_i \in \Gamma_2 \tag{3}$$

$$u_i = \delta; \quad x_i \in \Gamma_3 \tag{4}$$

which represents a generalized form of applied deformations that includes translations and rotations of the composite region. At the region of the inclusion-elastic medium interface which maintains contact we have

$$u_n = \mathbf{n} \cdot \mathbf{v} - (\mathbf{r}_0 \cdot \mathbf{s})\omega; \quad x_i \in \Gamma_4 \tag{5}$$

$$u_s = \mathbf{s} \cdot \mathbf{v} + (\mathbf{r}_0 \cdot \mathbf{n})\omega; \quad x_i \in \Gamma_4 \tag{6}$$

where $\mathbf{n}$ and $\mathbf{s}$ are respectively the unit directional vectors along the normal and tangential directions of $\partial\Omega_m$. The boundary conditions on Γ_5 will depend on the manner in which the non-linear phenomenon is characterized either in terms of Coulomb friction or interface plasticity. For example, in the case of a frictional response

$$u_n = \mathbf{n} \cdot \mathbf{v} - (\mathbf{r}_0 \cdot \mathbf{s})\omega; \quad x_i \in \Gamma_5 \tag{7}$$

$$t_s = \pm \mu t_n; \quad x_i \in \Gamma_5 \tag{8}$$

where t_s and t_n are respectively the shear and normal tractions and μ is the coefficient of friction between $\partial\Omega_m$ and $\partial\Omega_b$. Finally, since Ω_b is completely enclosed within Ω_m, all the contact forces acting on $\partial\Omega_b$ should satisfy global equilibrium. Therefore for a given state of deformation $\{v_x, v_y, \omega\}$ we have

$$\int_{\partial\Omega_b} \{t_n n_i + t_s s_i\} \, d\Gamma = 0; \quad (i = 1, 2) \tag{9}$$

$$\int_{\partial\Omega_b} \{(\mathbf{r}_0 \cdot \mathbf{s})t_n - (\mathbf{r}_0 \cdot \mathbf{n})t_s\} \, d\Gamma = 0 \tag{10}$$

where Eqs. (9) and (10), corresponding to the two traction resultants and moment resultant, can be rewritten in the form

$$\int_{\partial\Omega_b} [C]^T \{t\} \, d\Gamma = 0 \tag{11}$$

where $\{t\}$ are the traction vectors on $\partial\Omega_b$.

The boundary integral equation governing the displacements $u_l^{(i)}$ at the ith location in the isotropic linear elastic region Ω_m can be written in the form

$$\frac{1}{2}u_l^{(i)} + \int_\Gamma t_{lk}^* u_k \, d\Gamma = \int_\Gamma u_{lk}^* t_k \, d\Gamma; \quad (l, k = 1, 2) \tag{12}$$

where u_{lk}^* and t_{lk}^* are the fundamental solutions given in Sect. 5.7 (Appendix). A boundary element formulation of Eq. (12) will produce the following matrix equation

$$[H]\{U\} = [G]\{t\} \tag{13}$$

where $\{U\}$ and $\{t\}$ on Γ_4 and Γ_5 have been transformed as local variables $[u_n, u_s]$ and $[t_n, t_s]$ according to the transformation

$$\left[\begin{bmatrix} u_n \\ u_s \end{bmatrix}; \begin{bmatrix} t_n \\ t_s \end{bmatrix}\right] = \begin{bmatrix} n_1 & n_2 \\ s_1 & s_2 \end{bmatrix} \left[\begin{bmatrix} u_1 \\ u_2 \end{bmatrix}; \begin{bmatrix} t_1 \\ t_2 \end{bmatrix}\right]. \tag{14}$$

5.3 Frictional Interface

In this section we shall present a brief account of the incremental procedure adopted for the analysis of the interface frictional contact problem. The interface frictional contact involves dissipative phenomena which occur in boundary regions of unknown extent. In these circumstances the iterative procedure adopted in the analysis explicitly assumes that the final effect of a loading history can be represented as the cumulative effect of its increments. For example, the applied deformation δ (Fig. 1) can be represented as

$$\delta = \sum_{l=1}^{n} \Delta\delta^l = \delta^{n-1} + \Delta\delta^n \tag{15}$$

where n is the number of increments.

Similar definitions apply for other quantities such as u, t, ω, ..., etc. It must, however, be noted that there is no generalized proof of the uniqueness of the solution generated via the incremental formulation of frictional contact problems. Haslinger and Janovsky [46] indicate situations where stability of the frictional system cannot be established during the deformation history. Similar consideration apply in instances where the friction process is treated in a dynamic sense [47]. Despite these limitations the incremental procedure has been used quite successfully in the examination of special classes of problems involving interface friction phenomena [39–41]. Although the interfaces exhibit frictional dissipative phenomena, the region Ω_m displays elastic behaviour at any stage of deformation; consequently the boundary integral equation (12) is valid for incremental quantities. The appropriate matrix equation can be written as

$$[H]\{\Delta u^l\} = [G]\{\Delta t^l\}. \tag{16}$$

At each incremental stage, an iterative procedure must be employed to determine the extent of the slip region (Γ_5) and the frictionally locked region (Γ_4). For a Mohr-

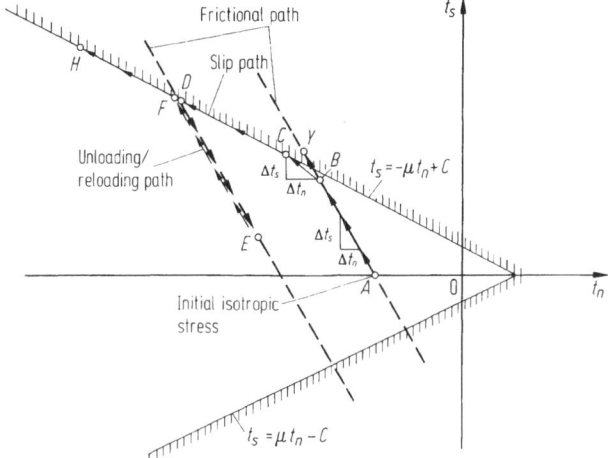

Fig. 2. Relation between normal and tangential tractions on $\partial\Omega_b$. Loading–unloading path A–B–C–D–E–F–H

Coulomb-type [48] interface contact, the relevant criteria used in the iterative process takes the form

$$|t_s| < \mu|t_n| + c; \quad x_i \in \Gamma_4 \tag{17}$$

and

$$|t_s| \not< \mu|t_n| + c; \quad x_i \in \Gamma_5 \tag{18}$$

where c is the interface cohesive strength. The Fig. 2 illustrates typical loading paths for the normal and tangential tractions on $\partial\Omega_b$. The point A in Fig. 2 corresponds to an initial isotropic stress state from which the frictional path AB originates. The boundary conditions for the incremental analysis are the following:

$$\Delta u^l = 0; \qquad x_i \in \Gamma_1 \tag{19}$$

$$\Delta t^l = 0; \qquad x_i \in \Gamma_2 \tag{20}$$

$$\Delta u^l = \Delta\delta^l; \quad x_i \in \Gamma_3 \tag{21}$$

$$\Delta u_n^l = \mathbf{n} \cdot \Delta\mathbf{v}^l - (\mathbf{r}_0 \cdot \mathbf{s})\Delta\omega^l; \quad x_i \in \Gamma_4 \tag{22}$$

$$\Delta u_s^l = \mathbf{s} \cdot \Delta\mathbf{v}^l + (\mathbf{r}_0 \cdot \mathbf{n})\Delta\omega^l; \quad x_i \in \Gamma_4. \tag{23}$$

The conditions (19)–(23) are valid until the location Y (Fig. 2) is reached. At Y, the condition (17) is no longer valid and it is necessary to re-derive the boundary conditions by focussing on the location C on the line $t_s = \pm\mu t_n \mp c$. Hence, the condition defined by Γ_5 should produce the result such that $t_s^l = \pm\mu t_n^l \mp c$ (see Fig. 3). Consequently

$$t_s^{l-1} + \Delta t_s^l = \pm\mu(t_n^{l-1} + \Delta t_n^l) \mp c \tag{24}$$

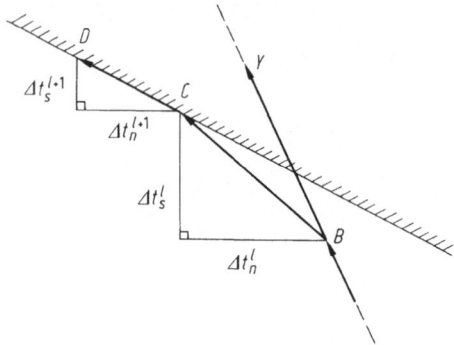

Fig. 3. The increments from frictional to slipping

and, assuming that no separation occurs at the inclusion-elastic medium interface, the requisite boundary conditions on Γ_s take the forms

$$\Delta u_n^l = \mathbf{n} \cdot \Delta \mathbf{v}^l - (\mathbf{r}_0 \cdot \mathbf{s}) \Delta \omega^l \tag{25}$$

$$\Delta t_s^l = \pm \mu \Delta t_n^l + R^l \tag{26}$$

where $R^l = (\pm \mu t_n^{l-1} - t_s^{l-1} \mp c)$. The sign of the coefficient of friction will be given by that of the sign of t_s when slipping occurs, i.e.,

$$\pm \mu = -\mu \operatorname{sign}(t_s). \tag{27}$$

When slipping has occurred at the $(l-1)^{\text{th}}$ increment, the residual traction R^l will be identically equal to zero. Writing $\{\Delta \mathbf{v}^l, \Delta \omega^l\}$ as ΔD^l and substituting the boundary conditions (19)–(26) we obtain the expression for the matrix equation (16):

$$[A^l]\{\Delta X^l\} = [B^l]\{\Delta D^l, \Delta \delta^l, 1\}. \tag{28}$$

Upon inversion, we obtain

$$\{\Delta X^l\} = [Q_1^l]\{\Delta D^l\} + [Q_2^l]\{\Delta \delta^l\} + [Q_3^l]. \tag{29}$$

The traction on the boundary $\partial \Omega_b$ can be written as

$$\{t^l\} = \{t^{l-1}\} + [C_1^l]\{\Delta X^l\}. \tag{30}$$

Using the equilibrium condition given by Eqs. (9), (30) can be re-written as

$$\int_{\partial \Omega_b} [C]^T [\{t^{l-1}\} + [C_1^l]\{\Delta X^l\}] \, d\Gamma = 0. \tag{31}$$

The resulting expression can be written as

$$[K^l]\{\Delta D^l\} + \{\Delta p^l\} = 0 \tag{32}$$

where

$$[K^l] = \int_{\partial \Omega_b} [C]^T [C_1^l][Q_1^l] \, d\Gamma. \tag{33}$$

In Eq. (32) $[K^l]$ may be considered as the stiffness matrix for the body Ω_b and $\{\Delta p^l\}$ takes the form

$$\{\Delta p^l\} = \int_{\partial\Omega_b} [C]^T [C_1^l][Q_2^l] \, d\Gamma \{\Delta\delta^l\}$$

$$+ \int_{\partial\Omega_b} [C]^T [[C_1^l][Q_3^l] + \{t^{l-1}\}] \, d\Gamma. \tag{34}$$

In summary, we can solve Eq. (32) and substitute the solution for $\{\Delta D^l\}$ for the boundary unknown $\{\Delta X^l\}$ and hence the cumulative responses of all dependent variables up to the lth stage.

If unloading occurs at a point D (see Fig. 2) then a negative value of the increment is implemented. The iteration processes is carried out to locate the recovery of the contact boundary on $\partial\Gamma_b$. The elastic unloading at the interface can be described by path DE (see Fig. 2) and the reloading path corresponds to EF.

5.4 Plastic Interface

The frictional modelling of the interface discussed in the previous section is one possible approach to the study of interface non-linear phenomena. An alternative to this technique would be to assume that the interface exhibits plastic phenomena. It is assumed that when an interface exhibits plastic response on $\partial\Omega_b$, its incremental deformation can be represented as a combination of the elastic and plastic displacement increments, i.e.,

$$\Delta u_n^l = \Delta u_{n(e)}^l + \Delta u_{n(p)}^l \tag{35}$$

$$\Delta u_s^l = \Delta u_{s(e)}^l + \Delta u_{s(p)}^l \tag{36}$$

where the subscripts (e) and (p) refer to the elastic and plastic components respectively. As before, we restrict attention to the class of problems in which there is no separation at the inclusion-elastic medium interface $\partial\Omega_b$. Consequently, the normal incremental displacement can still be described by Eq. (25). However, the slip displacement which occurs in the tangential direction due to the plastic flow can be summed as an additional tangential deformation as shown in Fig. 4. Therefore

$$\Delta u_s^l = \Delta\bar{u}_s^l + \Delta\lambda^l. \tag{37}$$

Also the boundary condition on Γ_4 will be a special case of that prescribed on Γ_5 when $\Delta\lambda^l = 0$. The boundary conditions on $\Gamma_4 + \Gamma_5 = \Gamma_6$ can be written as

$$\Delta u_n^l = n \cdot \Delta v^l - (r_0 \cdot s)\Delta\omega^l \tag{38}$$

$$\Delta u_s^l = s \cdot \Delta v^l + (r_0 \cdot n)\Delta\omega^l + \Delta\lambda^l. \tag{39}$$

Substituting the above boundary conditions together with those given on Γ_1, Γ_2 and Γ_3 into Eq. (16), we obtain the following:

$$\{\Delta X^l\} = [Q_1]\{\Delta D^l\} + [Q_2]\{\Delta\delta^l\} + [Q_3] + [L]\{\Delta\lambda^l\}. \tag{40}$$

It may be noted that the $[L]$ and $[Q]$ matrices are independent of the current stage of the increment l, which indicates that Eq. (40) can be applied for every increment.

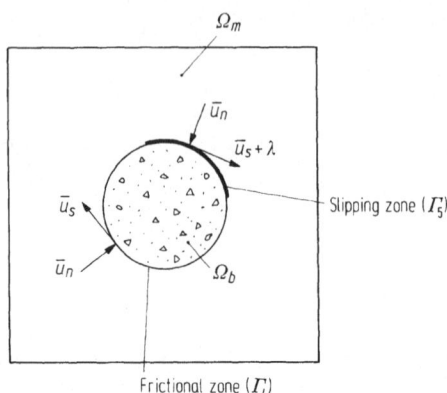

Fig. 4. The normal and tangential deformation on $\partial\Omega_b$

Following the equilibrium condition given by Eqs (9) and (10) we obtain a single equilibrium equation as

$$[K_{11}^l]\{\Delta\lambda^l\} + [K_{12}]\{\Delta D^l\} + [K_{13}]\{\Delta\delta^l\} + \{\Delta P_{11}\} + \{\Delta P_{12}^l\} = 0 \qquad (41)$$

where the matrices are defined by

$$[[K_{11}];[K_{12}];[K_{13}];\{\Delta P_{11}\}] = \int_{\partial\Omega_b} [C]^T[C_1][[L];[Q_1];[Q_2];[Q_3]]\,d\Gamma \qquad (42)$$

and

$$\{\Delta P_{12}^l\} = \int_{\partial\Omega_b} [C]^T\{t^{l-1}\}\,d\Gamma. \qquad (43)$$

Also $[K_{11}^l]$ is modified from $[K_{11}]$. Also one can define the normal and tangential traction of $\partial\Omega_b$ by the following

$$[\{\Delta t_n^l\};\{\Delta t_s^l\}] = [[C_n^l];[C_s^l]]\{\Delta X^l\} \qquad (44)$$

respectively. If the number of elements in which slipping occurs (i.e., yielding occurs) is m_s^l at the lth stage, there will be m_s^l extra unknowns in the Eqs. (41); namely $\{\Delta\lambda^l\}$. However, there will be m_s^l conditions pertaining slipping which need to be satisfied by the tractions $\{\Delta t_n^l\}$ and $\{\Delta t_s^l\}$. The resulting m_s^l equations can be expressed in a matrix form as

$$[K_{21}^l]\{\Delta\lambda^l\} + [K_{22}^l]\{\Delta D^l\} + [K_{23}^l]\{\Delta\delta^l\} + \{\Delta P_{21}^l\} = \{\Delta P_{22}^l\} \qquad (45)$$

where the matrices are given by

$$[[K_{21}^l];[K_{22}^l];[K_{23}^l];\{\Delta P_{21}^l\}] = [[C_s^l]\mp\mu[C_n^l]][[L];[Q_1];[Q_2];[Q_3]] \qquad (46)$$

and the vector on the right hand side of Eq. (45) is

$$\{\Delta P_{22}^l\} = \{R^l\}. \qquad (47)$$

The sign for $\mp\mu$ in (46) should be assigned according to every individual slipping element. Together with the equation given by Eq. (41), the complete system of equations for $[\{\Delta\lambda^l\}, \{\Delta D^l\}]$ can be written as

$$\begin{bmatrix} K_{11} & K_{12} \\ K_{21} & K_{22} \end{bmatrix} \begin{bmatrix} \Delta\lambda^l \\ \Delta D^l \end{bmatrix} + \begin{bmatrix} \Delta P_1^l \\ \Delta P_2^l \end{bmatrix} = 0. \tag{48}$$

The equation (48) represents a system of equations of order $(m_s^l + 3)$ depending on the number of elements which undergo yielding. Upon solution of Eq. (48), the incremental solution $\{\Delta X^l\}$ can be obtained from Eq. (40).

The interface plasticity approach to the problem gives a single type of boundary condition on Γ_6 or $\partial\Omega_b$. Consequently the boundary element matrix equation is solved only once. The iteration takes place over the values of $\{\Delta\lambda^l\}$ rather than on the location of the boundary between Γ_4 and Γ_5 (i.e., $\Gamma_4 \cap \Gamma_5$). The equation for $\{\Delta X^l\}$ given in Eq. (40) can be re-employed for any number of steps (l) in the incremental analysis.

5.5 Numerical Results

The numerical procedures outlined in the previous sections have been applied to the study of the influence of non-linear interface response on the plane-strain shear behaviour of the composite region. The composite consists of a square region of elastic material (of dimension $2a$) which is bounded internally by a rigid inclusion of circular cross section (of diameter $2b$). This particular problem serves as an effective cell model which can be employed in the study of non-linear phenomena which can occur at the interface between a fibre and matrix of a fibre-reinforced material. The Fig. 5 shows the boundary element idealization of the quadrant of the domain. The entire composite is subjected to an initial isotropic confining stress σ_0, to prevent separation at the inclusion-elastic medium interface. The numerical results presented in this section are purely the purposes of illustration of the numerical schemes employed in the analysis of non-linear interface phenomena. For this reason the relative dimensions of the composite region are assigned the following specific values; $a = 2.0$; $b = 1.5$. Altogether 28 constant boundary elements are

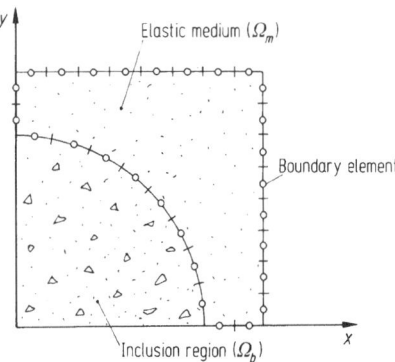

Fig. 5. The boundary element discrete for the composite region

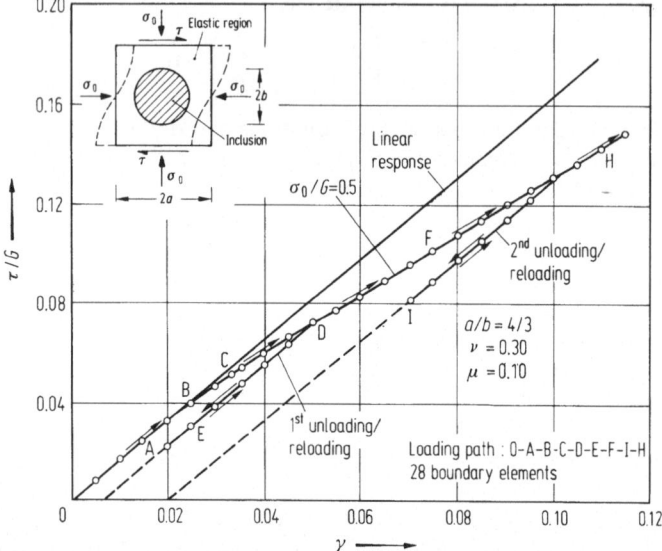

Fig. 6. The shear stress-strain response of the composite using frictional interface

used to model the quadrant region (Fig. 5). The results presented in Figs. 6 and 7 deal with situation where the composite region is subjected to direct shear loading as shown in the inset in Fig. 6. The shear strain in the composite region is defined as $\gamma = \tan^{-1}(\Delta^*/b)$ where Δ^* is the prescribed shear displacement at the surfaces Γ_1 and Γ_3. The Fig. 6 shows the normalized shear stress (τ/G) vs. shear strain for the case where the interface non-linearities are examined via the interface friction algorithm. The results in Fig. 6 illustrate the manner in which the frictional response at the interface constributes to the development of a global non-linear response in the composite region. The Fig. 7 shows the locations of the frictional and slip zones at the inclusion-elastic medium interface. The Fig. 8 shows equivalent results derived for the case where the outer boundary of the composite region is subjected to a deformation which corresponds to a state of pure shear. In this case the method of analysis of the interface non-linear phenomenon employs the interface plasticity algorithm. The results are in general agreement with equivalent results derived via the interface friction algorithm. The results presented in Fig. 9 illustrate the location of the slip/friction interface region at different stages of the loading/unloading process. The results presented in Fig. 10 focus on the interface non-linearity problem for the situation where interface frictional slip occurs in the unloading mode. The analyses is performed by invoking the interface plasticity algorithm. These results focus on both the Coulomb friction model and the Mohr-Coulomb model which incorporates interface cohesion. The results indicate the typical hystersis phenomena associated with materials which possess dissipative phenomena of the plasticity/frictional type.

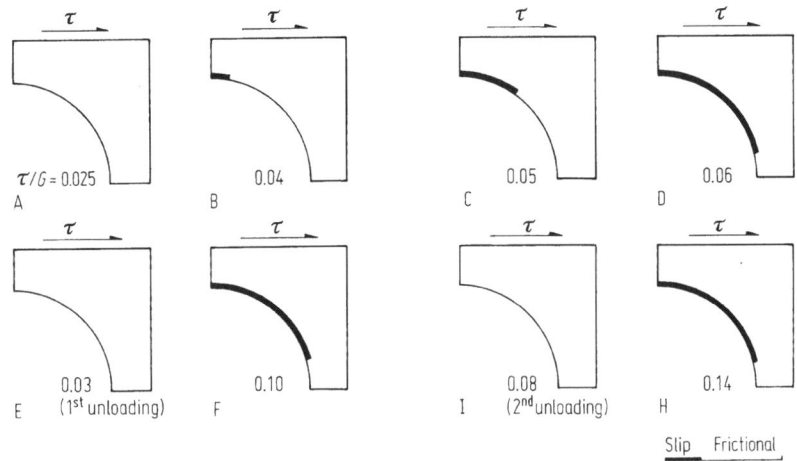

Fig. 7A–H. Locations of the frictional and slip zone during the loading history. ($\sigma_0/G = 0.5$): simple shear response (see inset Fig. 6)

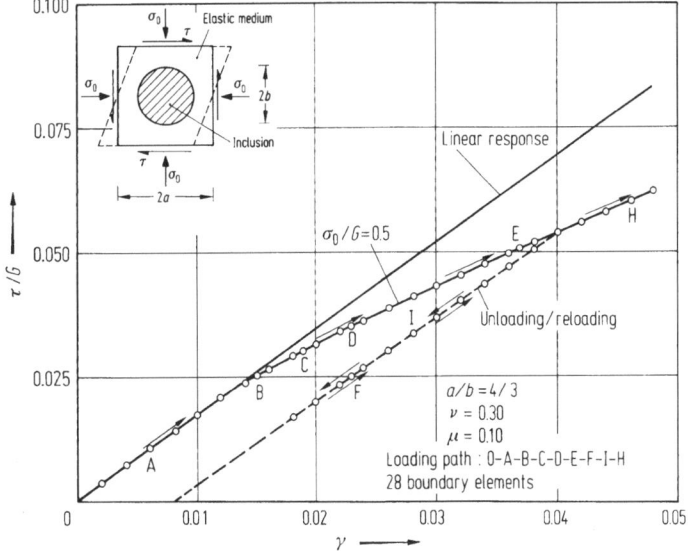

Fig. 8. The shear stress-strain response of the composite using plastic interface

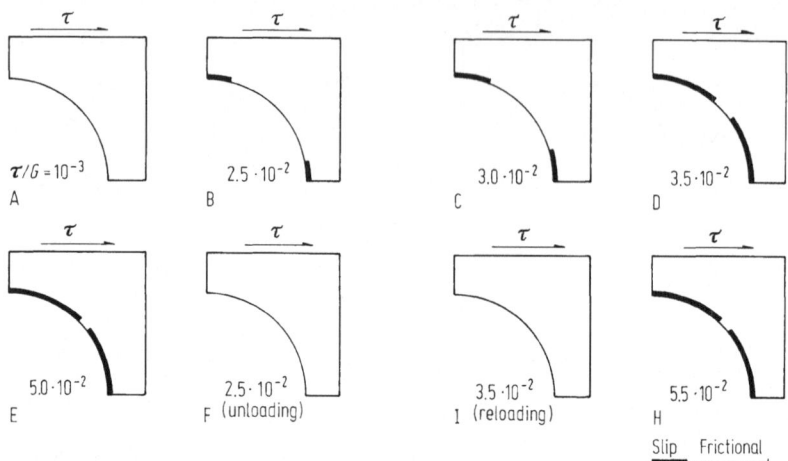

Fig. 9A–H. Locations of the frictional and the slip zone: pure shear response (see inset Fig. 8)

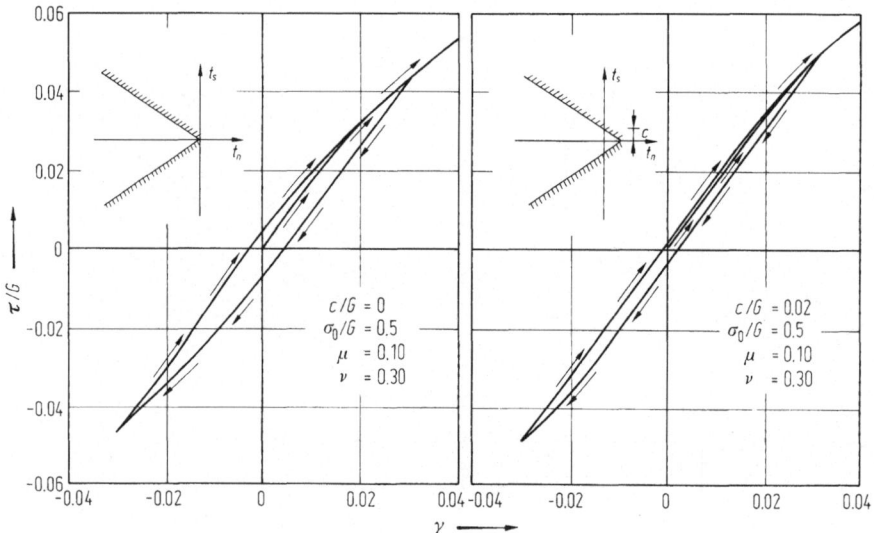

Fig. 10. Shear stress vs shear strain response of the composite

5.6 Conclusions

The non-linear phenomena which occur at the interfaces of material bodies can exert a significant influence on the overall contact response of the material regions. Such non-linear phenomena can display varied responses which range from Coulomb friction and finite friction to models with arbitrary non-linear responses. These latter responses may be derived from experimental results or micro-mechanical considerations. The theoretical analyses which incorporate these models are quite complicated and recourse must be made to the use of numerical methods of stress analysis in order to examine contact problems of engineering interest. The present article considers the application of boundary element schemes for the study of interface non-linear phenomena. The numerical basis for the boundary element analysis of non-linear interface phenomena can be developed by employing either an interface frictional response or an interface plasticity model. The present article employs both approaches to examine the plane strain shear response of a square elastic region which is bounded internally by a rigid circular inclusion; the interface between the inclusion and the elastic medium exhibits a non-linear response. The composite region is subjected to an initial isotropic stress which maintains contact between the elastic region and the rigid inclusion. The numerical studies conducted indicate that both the interface friction model and the interface plasticity model can be employed quite effectively for the incremental stress analysis of the shear response of the composite. Specific results presented in the paper illustrate the manner in which local non-linearities at material interfaces can contribute to global non-linear phenomena in the overall responses of the composite region. The numerical schemes developed in connection with these studies assume that there is no separation at the matrix-inclusion interface. The basic methodologies presented in the article can however, be extended to include separation at the inclusion-elastic medium interface and generalized elastic-plastic type interface non-linear phenomena.

5.7 Appendix

The fundamental solutions for u_{lk}^* and P_{lk}^* referred to in the text are given by (see Ref. [42])

$$u_{lk}^* = \frac{1}{8\pi G(1-v)} \left[(3-4v)\ln\left(\frac{1}{r}\right)\delta_{lk} + r_{,l}r_{,k} \right] \tag{A1}$$

and

$$P_{lk}^* = -\frac{1}{4\pi(1-v)r}\{[(1-2v)\delta_{lk} + 2r_{,l}r_{,k}]r_{,n}$$
$$- (1-2v)[r_{,l}n_k - r_{,k}n_l]\}. \tag{A2}$$

The Eqs. (A1) and (A2) are applicable to the plane strain problem. Equivalent expressions for the plane stress problem can be obtained by replacing v in (A1) and (A2) by $v/(1+v)$.

5.8 Notations

Ω_M	Elastic medium	$\partial\Omega_b$	Boundary of the rigid inclusion
Ω_b	Rigid inclusion	x_i	Co-ordinates system ($i = 1, 2$)
$\partial\Omega_m$	Boundary of the elastic medium	u	Displacement

t	Traction	Γ_2	Boundary with prescribed traction
$\bar{u}_0$	Prescribed boundary displacement	Γ_3	Boundary with prescribed displacement (applied)
$\bar{t}_0$	Prescribed boundary traction	Γ_4	Perfectly bonded boundary
δ	Prescribed displacement	Γ_5	Boundary with slip
n	Unit outward normal vector	$[G]$	Matrix of displacement coefficients
s	Unit tangential vector		
v	Translation of rigid body	$[H]$	Matrix of traction coefficients
ω	Rotation of rigid body		
σ_{ij}	Stress field in Ω_m	$\{\Delta X^l\}$	Boundary unknown vector at the lth increment
G	Linear elastic shear modulus		
v	Poisson's ratio	m_s^l	Number of elements on the boundary with slip; at the lth increment
μ	Coefficient of friction between $\partial\Omega_m$ and $\partial\Omega_b$		
γ	Shear strain	$[K_{ij}^l]$	Stiffness matrices at the lth increment
τ	Shear stress		
σ_0	Initial normal stress	u_{lk}^*	Fundamental solutions for displacements $(l, k = 1, 2)$
$\Delta\delta^l$	Prescribed displacement at l^{th} increment		
u_n	Normal component of the displacement	t_{lk}^*	Fundamental solutions for traction $(l, k = 1, 2)$
u_s	Tangential component of the displacement	n	Number of increments
		Γ_6	$\Gamma_4 \cup \Gamma_5$
$\Delta\lambda^l$	Tangential component of plastic flow at the lth increment	c	Adhesion between $\partial\Omega_m$ and $\partial\Omega_b$
		$2a$	Side dimension of square region
Γ_1	Boundary with prescribed displacement	$2b$	Diameter of rigid inclusion

References

1 Duvaut, G. and Lions, J.L., *Inequalities in Mechanics and Physics*, Springer Verlag, Berlin (1975).
2 Gudehus, G., (Ed.), *Finite Elements in Geomechanics*, John Wiley, New York, (1977).
3 Desai, C.S. and Christian, J.T., (Eds.), *Numerical Methods in Geomechanics*, McGraw-Hill, New York, (1977).
4 de Pater, A.D. and Kalker, J.J., (Eds.)., *The Mechanics of Contact Between Deformable Media*, Proc. IUTAM Symposium, Enschede, Deflt Univ. Press, Delft (1975).
5 Selvadurai, A.P.S., *Elastic Analysis of Soil-Foundation Interaction, Developments in Geotechnical Engineering*, Vol. 17, Elsevier Scientific Publ. Co., Amsterdam, (1979).
6 Gladwell, G.M.L., *Contact Problems in the Classical Theory of Elasticity*, Sijthoff and Noordhoft, The Netherlands, (1980).
7 Kikuchi, N. and Oden, J.T., *Contact Problems in Elasticity*, Texas Institute for Computational Mechanics, Rept. 79-8, The University of Texas, Austin, (1979).
8 Panagiotopoulos, P.D., *Inequality Problems in Mechanics and Applications*, Birkhauser Verlag, Basel (1985).
9 Johnson, K.L., *Contact Mechanics*, Cambridge University Press, U.K., (1985).
10 Selvadurai, A.P.S. and Voyiadijs, G.Z. (Eds.), *Mechanics of Material Interfaces*, Studies in Applied Mechanics, Vol. 11, Elsevier Scientific Publ. Co., Amsterdam, (1986).

11 Bowden, F.P. and Tabor, D., *Friction and Lubrication of Solids*, Vols. I and II, Oxford University Press, Oxford, (1964).

12 Ufliand, Ia. S., *Survey of Articles on the Application of Integral Transforms in the Theory of Elasticity*, English Transl. Sneddon, I.N. (Ed.), Raleigh, North Carolina State University, (1965).

13 Galin, L.A., *Contact Problems in the Theory of Elasticity*, English Transl., Sneddon, Z.N. (Ed.), Raleigh, North Carolina State University, (1961).

14 Dundurs, J. and Stippes, M., 'Role of elastic constants in certain contact problems', *J. Appl. Mech.*, *37*, pp. 965–970 (1970).

15 Sneddon, I.N., *Application of Integral Transforms in The Theory of Elasticity*, CISM Courses and Lectures No. 220, Springer Verlag, Wein, (1975).

16 Spence, D., 'Self similar solutions to adhesive contact problems', *Proc. Roy. Soc., Ser. A.*, *305*, pp. 55–80 (1968).

17 Kalker, J.J., 'Variational principles of contact elastostatics', *J. Inst. Math. Applics.*, *20*, pp. 199–219 (1977).

18 Turner, J.R., 'The frictional unloading problem on a linear elastic halfspace', *J. Inst. Math. Applics.*, *24*, pp. 439–469 (1979).

19 Goodman, R.E., Taylor, R.L. and Brekke, T.L., 'A model for the mechanics of jointed rock', *J. Soil Mech. Fdn. Div. Proc. ASCE*, *94*, pp. 637–659 (1968).

20 Goodman, R.E. and Dubois, J., 'Duplications of dilatancy in analysis of jointed rocks', *J. Soil Mech. Fdn. Div., Proc. ASCE*, *98*, pp. 399–422 (1972).

21 Ghabbousi, J., Wilson, E.L. and Isenberg, J., 'Finite Elements for rock joints and interfaces', *J. Soil Mech. Fdn. Div. Proc. ASCE*, *99*, pp. 833–848 (1973).

22 Zienkiewicz, O.C., Best, B., Dullage, C. and Stagg, K., 'Analysis of non-linear problems in rock mechanics with particular reference to jointed rock systems', *Proc. 2nd Congress Int. Soc. Rock Mech.*, Belgrade, Yugoslavia, *3*, pp. 501–509 (1970).

23 Fredriksson, G., 'Finite element solution of surface non-linearities in structural mechanics', *Computers and Structures*, *6*, pp. 281–290 (1976).

24 Boulon, M., Darve, F., Desrues, J. and Foray, P., 'Soil-structure coupling, nonlinear rheological relationships and boundary conditions in soil mechanics', *Computers and Structures*, *9*, pp. 293–303 (1978).

25 Pande, G.N. and Sharma, K.G., 'On joint/interface elements and associated problems of numerical ill-conditioning', *Int. J. Num. Anal. Meth. Geomech.*, *3*, pp. 293–300 (1979).

26 Ito, Y.M., England, R.H. and Nelson, R.B., 'Computational methods for soil-structure interaction problems', *Computers and Structures*, *13*, pp. 157–162 (1981).

27 Selvadurai, A.P.S. and Faruque, M.O., 'The influence of interface friction on the performance of cable jacking tests of rock masses', *Proc. Impl. Comp. Procedures and Stress-Strain Laws in Geotech Eng.*, Desai, C.S. and Saxena, S.K., Eds., Illinois, Runcorn Press, 1, pp. 169–183 (1981).

28 Desai, C.S., 'Behaviour of interfaces between structural and geologic media', *Proc. Int. Conf. on Recent Advances in Geotech. Earthq. Eng. and Soil Dynamics*, (Prakash, S., Ed.), St. Louis, Mo., *2*, pp. 619–638, (1981).

29 Desai, C.S. and Nagaraj, B.K., 'Constitutive modelling for interfaces under cyclic loading' in *Mechanics of Material Interfaces*, (Selvadurai, A.P.S. and Vorjiadjis, G.Z., Eds.), Vol. II, Studies in Applied Mechanics, Elsevier Scientific Publ. Co., pp. 97–108, (1986).

30 Goodman, R.E., 'Analysis of jointed rocks', Ch. 11 in *Finite Elements in Geomechanics*, (Gudehus, G., Ed.), John Wiley, New York, pp. 351–376 (1977).

31 Desai, C.S., 'Soil-structure interaction and simulation problems', Ch. 7 in *Finite Elements in Geomechanics*, (Gudehus, G., Ed.), John Wiley, New York, pp. 209–250 (1977).

32 Herrmann, L.R., 'Finite element analysis of contact problems', *J. Eng. Mech. Div. Proc. ASCE, 104*, 1042–1057 (1978).

33 Wilson, E.L., 'Finite elements for foundations, joints and fluids', Ch. 10 in *Finite Elements in Geomechanics*, (Gudehus, G., Ed.), John Wiley, New York, pp. 319–350 (1977).

34 Gaertner, R., 'Investigation of plane elastic contact allowing for friction', *Computers and Structures*, *7*, pp. 59–63 (1977).

35 Okamoto, N. and Nabazawa, M., 'Finite element incremental contact analysis with various frictional conditions', *Int. J. Num. Meth. Eng.*, *14*, pp. 337–357 (1979).

36 Zienkiewicz, O.C., *The Finite Element Method in Engineering Science*, McGraw-Hill, New York (1978).

37 Andersson, T., 'The boundary element method applied to two-dimensional contact problems with friction', in *Boundary Element Methods*, Proc. 3rd Int. Seminar, Irvine, California (Brebbia, C.A., Ed.), Springer-Verlag, Berlin (1981).

38 Andersson, T., 'The second generation boundary element contact program' in *Boundary Element Methods in Engineering*, Proc. 4th Int. Seminar, Southampton, U.K. (Brebbia, C.A. Ed.) Springer-Verlag, Berlin (1982).

39 Andersson, T., and Allan Persson, B.G., 'The boundary element method applied to two-dimensional contact problems', Ch. 5 in *Progress in Boundary Element Methods*, (Brebbia, C.A., Ed.), CML Publications, U.K., 2, pp. 136–157 (1983).

40 Paris, F. and Garrido, J.A., 'On the use of discontinuous elements in two-dimensional contact problems', *Proc. 7th Boundary Element Conference*, (Brebbia, C.A. and Maier, G., Eds.) Como, Springer-Verlag, Berlin, pp. 13.27–13.39 (1985).

41 Selvadurai, A.P.S. and Au., M.C., 'Response of inclusions with interface separation, friction and slip', *Proc. 7th Boundary Element Conference* (Brebbia, C.A. and Maier, G., Eds.), Como, Springer-Verlag, Berlin, pp. 14.109–14.127 (1985).

42 Brebbia, C., *The Boundary Element Method for Engineers*, Pentech Press, 2nd. ed., (1980).

43 Telles, J.C.F., '*The Boundary Element Method Applied to Inelastic Problems*', Lecture notes in Engineering, 1, Springer-Verlag, Berlin, (1983).

44 Banerjee, P.K. and Butterfield, R., '*Boundary Element Methods in Engineering Science*', McGraw-Hill, New York, (1981).

45 Brebbia, C.A., Tells, J.C.P. and Wrobel, L.C., '*Boundary Element Techniques, Theory and Applications in Engineering*', Springer-Verlag, (1984).

46 Haslinger, J. and Janovsky, V., 'Contact problems with friction', in *Trends in Application of Pure Mathematics to Mechanics*, (Brilla, J., Ed.) IV, pp. 74–100 (1983).

47 Martins, J.A.C. and Oden, J.T., 'Interface models, variational principles and numerical solutions for dynamic friction problems', in *Mechanics of Material Interfaces*, (Selvadurai, A.P.S. and Voyiadjis, G.Z., Eds.), Studies in Applied Mechanics, Vol. II, Elsevier Scientific Publ. Co., pp. 3–22 (1986).

48 Chen, W.-F., *Limit Analysis and Soil Plasticity*, Developments in Geotechnical Engineering, Vol. 7, Elsevier Scientific Publ. Co., Amsterdam, (1975).

Chapter 6

Heterogeneities in Flows Through Porous Media by the Boundary Element Method

by A.H.-D. Cheng

6.1 Introduction

Of late, the Boundary Element Method (BEM) has emerged as a serious contender in the computational mechanics arena. It has been successfully applied to numerous engineering problems. The general perception of the BEM is that it is inherently more efficient because it is a boundary, instead of a domain, scheme. It is also probably more accurate because Green's function, which is admissible to the governing equation, is used as weighing functions. The major shortcoming of the BEM, however, is its relative inability in dealing with nonlinearities and heterogeneities, which are frequently encountered in engineering. To overcome the difficulties, some authors introduced variations, in which domain discretization and/or integration are usually needed, of the genuine boundary method. Although ways have been devised to optimize the domain treatment, the inherent efficiency of the boundary method can be seriously hurt.

It is the purpose of this article to review and to present innovative ideas to deal with heterogeneities in flows through porous media. It is not intended to be an ultimate solution of heterogeneities using the Boundary Element technique. We hope to attract researchers' attention to this potentially rich area of research.

We organize the sections as follows. In Sect. 6.2, the governing equations for Darcy's flow are reviewed. Medium and fluid heterogeneities, respectively, are treated in Sects. 6.3 and 6.4. Section 6.5 is devoted to embedded heterogeneities, which is then followed by concluding remarks.

6.2 Governing Equations for Darcy's Flow

The equation of motion for flows through saturated, nondeformable porous media is given by the generalized Darcy's law

$$\frac{\mu}{k} \boldsymbol{q} = -\nabla p - \rho g \nabla z \tag{1}$$

where $\boldsymbol{q}$ = specific discharge vector; p = pressure; μ = dynamic viscosity of fluid; ρ = density of fluid; k = intrinsic permeability; g = acceleration of gravity; and z = Cartesian coordinate measured in the opposite direction of the gravity. The continuity requires that

$$\frac{\partial \rho}{\partial t} + \nabla \cdot (\rho \boldsymbol{q}) = 0 \tag{2}$$

where t = time. The equation for an incompressible, heterogeneous fluid is

$$\frac{D\rho}{Dt} = \frac{\partial \rho}{\partial t} + \boldsymbol{q} \cdot \nabla \rho = 0. \tag{3}$$

For a fluid at constant temperature we assume that the viscosity remains constant when fluid particles are traced

$$\frac{D\mu}{Dt} = \frac{\partial \mu}{\partial t} + \boldsymbol{q} \cdot \nabla \mu = 0. \tag{4}$$

We ignore herein, for simplicity, the diffusional and the dispersive mechanisms. Combination of Eq. (2) and Eq. (3) yields the continuity equation for incompressible fluid flow

$$\nabla \cdot \boldsymbol{q} = 0. \tag{5}$$

In the general flow problems with negligible mass and heat transfer, Eqs. (1), (3), (4) and (5) must be solved simultaneously, in which k is a function of space, and ρ and μ are functions of both space and time. For the case of homogeneous fluid, significant simplification of the governing equation results. Taking divergence of Eq. (1) and using Eq. (5) give

$$\nabla \cdot (K \nabla H) = 0 \tag{6}$$

in which $K = \rho g k/\mu$ = hydraulic conductivity; and $H = p/\rho g + z$ = piezometric head. Further, if the porous medium is homogeneous, i.e., K = constant, Eq. (6) is reduced to the Laplace equation

$$\nabla^2 H = 0 \tag{7}$$

which is one of the most studied equations in engineering.

6.3 Heterogeneous Porous Media

In this section we deal with Eq. (6). Several Boundary Element approaches are discussed in the following.

6.3.1 Constant Hydraulic Conductivity

The material is homogeneous in this case and K is a constant. The governing equation is the Laplace Eq. (7) and the application of the BEM is most straight-forward. The boundary integral equation is (see, e.g., Ref. [19])

$$\alpha H = \int \left(G^* \frac{\partial H}{\partial n} - H \frac{\partial G^*}{\partial n} \right) dS \tag{8}$$

where n = outward normal of the boundary; dS denotes integration with respect to the boundary, S; G^* is free-space Green's function of the Laplace equation

$$\nabla^2 G^* = - \begin{Bmatrix} 2\pi \\ 4\pi \end{Bmatrix} \delta(\bar{x}; \bar{x}') \tag{9}$$

in which $\delta(\bar{x}; \bar{x}')$ = Dirac delta function; the upper and lower quantities in the braces are for two and three-dimensional problems, respectively; and corresponding Green's functions of the above are $G^* = -\ln r$ and $1/r$, respectively. The constant factor α in Eq. (8) is determined from the Cauchy principal value integration of the kernal singularity.

The numerical technique solving Eq. (8) is similar to that of the FEM which includes discretization of the geometry into "boundary elements", polynomial approximation of the field variables, integration by numerical quadratures, etc. The details of the implementation are widely reported in the literatures, thus are not repeated herein.

6.3.2 Piecewise Homogeneous Media

6.3.2.1 Multizone Formulation

In groundwater flow the permeability is often specified as constant over certain regions or zones. There may be several zones in a specific problem. For those problems, the boundary integral equation Eq. (8) can be applied to each zone and a matrix equation is obtained with each application. The matrices are then combined and condensed using the interzonal compatibility conditions, namely that the flux and the piezometric head must be continuous

$$H_1 = H_2 \tag{10a}$$

$$-K_1 \left(\frac{\partial H}{\partial n} \right)_1 = K_2 \left(\frac{\partial H}{\partial n} \right)_2 = q_n \tag{10b}$$

where the subscripts 1 and 2 denote parameters associated with the two adjacent zones; and q_n = the normal flux across the interface. The result is a linear system

$$[A]\{x\} = \{b\} \tag{11}$$

in which $[A]$ is an $(M + 2N) \times (M + 2N)$ matrix; M = the number of external boundary nodes; and N = the number of interzonal nodes.

For multizone problems, the resulting matrix is not fully populated, in which a significant portion is occupied by blocks of zero entities. Similar matrices are produced in the combination of BEM with FEM [32]. The matrices, however, are unsymmetric and non-banded. A direct elimination algorithm is costly, rendering it impractical. To overcome such difficulty, Lachat and Watson [15] devised a scheme which optimally numbers the subdomain nodes. An out-of-core, banded solver is then used for the solution. Lafe et al. [16] sought the Finite Element idea of substructuring to reduce the size of the matrix. Crotty [9] used a block-based Gaussian elimination routine which ignores the zero blocks. Bettess [1] tested an algorithm which uses a direct solution scheme for the diagonal band and an interative scheme for the smaller off-band terms. All of the above schemes claimed some degree of success.

It seems that for problems with large numbers of zones, say, of the order 10 or above, the success of the BEM is closely tied to its ability in attacking such matrices efficiently. Since we have achieved only limited success in the solution algorithm for this type of linear system, the multizone BEM approach is so far not very practical for highly heterogeneous material, which calls for the division into large number of piecewise homogeneous zones.

6.3.2.2 Single Potential Formulation

In the zoned media, the potential is continuous across the interfaces. This gives us the opportunity to define a single potential for the entire domain. Consequently, a scheme leading to the reduction of the size of the final matrix can be achieved.

To construct the boundary integral equation valid for the entire domain, we view the zoned media as a continuously varied material. The governing equation is Eq. (6) with $K = K(x, y, z)$. The following integral equation, as derived in Ref. [3], can be utilized

$$\alpha KH = \int \left(KG^* \frac{\partial H}{\partial n} - KH \frac{\partial G^*}{\partial n} \right) dS + \int (H \nabla G^* \cdot \nabla K) dV \tag{12}$$

where dV represents the volume integral. For the piecewise homogeneous media, the volume integral is zero everywhere except at the interfaces where finite jumps of hydraulic conductivity are observed. To evaluate the volume integral at the interfaces we follow the idea of Ref. [10] to approximate the interzonal boundary as a transition region of finite width Δn in which K changes continuously from K_1 to K_2 in the direction normal to the surface and remains constant in the direction tangential to the surface. In the limit of Δn approaching zero we obtain the following

$$\int (H \nabla G^* \cdot \nabla K) dV = \lim_{\Delta n \to 0} \int dS_i \int_0^{\Delta n} H \frac{\partial G^*}{\partial n} \frac{\partial K}{\partial n} dn$$

$$= \lim_{\Delta n \to 0} \int H \frac{\partial G^*}{\partial n} dS_i \int_0^{\Delta n} \frac{\partial K}{\partial n} dn$$

$$= \int H \frac{\partial G^*}{\partial n} (K_2 - K_1) dS_i \tag{13}$$

where $S_i = $ the interzonal boundary. Equation (12) now involves only surface integrals

$$\alpha KH = \int \left(KG^* \frac{\partial H}{\partial n} - KH \frac{\partial G^*}{\partial n} \right) dS + \int H \frac{\partial G^*}{\partial n} (K_2 - K_1) dS_i \tag{14}$$

where S is the external boundary.

Equation (14) can be applied using both the internal and the external nodes as the base points. It is to be noted, however, when a base point is located at a smooth interzonal boundary, the left side of Eq. (14) should be modified to $\alpha(K_1 + K_2)H/2$ to correctly take into account the Cauchy principal value integration. Since only the potential H is formulated on the interzonal boundary, taking M nodes on the external boundary and N nodes on the internal boundary yields an $(M + N) \times$

$(M + N)$ system solving for the $M + N$ unknowns. As compared to the $M + 2N$ unknowns in the multizone and multipotential approach, a reduction in matrix size is achieved.

After the solution, if the flux across the interzonal boundary is also to be evaluated, the following equation, which is obtained by differentiating Eq. (14), can be used

$$\alpha q_n(\bar{x}) = \int \left[K(\bar{x}') \frac{\partial G^*(\bar{x}';\bar{x})}{\partial n(\bar{x})} \frac{\partial H(\bar{x}')}{\partial n(\bar{x}')} - K(\bar{x}')H(\bar{x}') \frac{\partial^2 G^*(\bar{x}';\bar{x})}{\partial n(\bar{x})\partial n(\bar{x}')} \right] dS(\bar{x}')$$

$$+ \int H(\bar{x}') \frac{\partial^2 G^*(\bar{x}';\bar{x})}{\partial n(\bar{x})\partial n(\bar{x}')} [K_2(\bar{x}') - K_1(\bar{x}')] dS_i(\bar{x}') \tag{15}$$

where $\bar{x}$ is the base point and $\bar{x}'$ is the field point.

Wei and Liggett [30] devised a similar scheme in which interzonal flux, instead of interzonal potential, is directly solved. For base points located at the external and the internal boundaries, respectively, the following integral equations are used

$$\alpha H = \int \left(G^* \frac{\partial H}{\partial n} - H \frac{\partial G^*}{\partial n} \right) dS + \int G^* q_n \left(\frac{1}{K_1} - \frac{1}{K_2} \right) dS_i \tag{16}$$

and

$$\alpha \left[\frac{1}{K_1(\bar{x})} + \frac{1}{K_2(\bar{x})} \right] q_n(\bar{x}) = \int \left[\frac{\partial G^*(\bar{x}';\bar{x})}{\partial n(\bar{x})} \frac{\partial H(\bar{x}')}{\partial n(\bar{x}')} - H(\bar{x}') \frac{\partial^2 G^*(\bar{x}';\bar{x})}{\partial n(\bar{x})\partial n(\bar{x}')} \right] dS(\bar{x}')$$

$$+ \frac{\partial G^*(\bar{x}';\bar{x})}{\partial n(\bar{x})} q_n(\bar{x}') \left[\frac{1}{K_1(\bar{x}')} - \frac{1}{K_2(\bar{x}')} \right] dS_i(\bar{x}'). \tag{17}$$

The size of the solution matrix is the same as that based on Eq. (14). A two-dimensional, three-zone problem was solved in [30].

6.3.3 Variable Hydraulic Conductivity

In the case of a continuously varied hydraulic conductivity, the governing equation is (6) with $K = K(x, y, z)$. This problem can be attacked using either Green's function for the Laplace equation or direct Green's function of Eq. (6). The formulations and tradeoffs of each approach are discussed in the following.

6.3.3.1 Iterative Scheme

The idea presented herein is similar to those employed for material nonlinearities using the Boundary Element technique. Typically, Green's functions for linear, homogeneous equations are adopted. The tradeoffs are that volume integrals and solution iterations become necessary [16, 23].

For the current problem we rearrange Eq. (6) to obtain the Poisson type equation

$$\nabla^2 H = \frac{1}{K} \nabla K \cdot \nabla H. \tag{18}$$

The boundary integral equation for the above is

$$\alpha H = \int \left(G^* \frac{\partial H}{\partial n} - H \frac{\partial G^*}{\partial n} \right) dS + \int \left(\frac{G^*}{K} \nabla K \cdot \nabla H \right) dV \tag{19}$$

where we have utilized Green's function for Laplace equation, G^*. The above equation can be solved as a Poisson equation if iterative scheme is used to update the value of H in the volume integral. A similar equation is Eq. (12), where the potential, instead of the gradient of the potential, is formulated in the volume integral. Equation (12) appears to be easier to implement.

We describe the procedure solving Eq. (12) as follows:

a) The domain is subdivided into internal cells similar in geometry to the Finite Element cells.

b) For the initial trial, Eq. (8) is first solved. Internal solutions are then found at the pivot points of the internal cells by numerical integration.

c) Equation (12) is discretized and numerically integrated. The domain integral, which contributes only to the right hand side of the matrix equation, is performed using the trial values obtained in the previous step.

d) The Boundary Element matrix is inverted (or decomposed) and the boundary solution is found from matrix multiplication.

e) The potentials at the interior quadrature points are updated by numerically integrating Eq. (12) using the new boundary values.

f) The left side of the matrix equation depends only on the geometry of the boundary mesh and the local permeability. It remains unchanged in the iterative process. The volume integral contributes to the right hand side and must be updated using the information obtained in step e).

g) Further solution of the matrix equation to obtain boundary solution involves only matrix multiplication and no inversion or decomposition.

h) Repeat steps e) through g) until convergence is achieved.

6.3.3.2 Domain Scheme

We briefly describe the domain scheme as proposed in Ref. [18]. This scheme is based upon the same integral equation of the iterative scheme, e.g., Eq. (19). Interior mesh and domain integration are also required. The two methods depart in the treatment of the potential at the interior nodes. The domain method formally treats the interior potentials as unknowns. Base points are distributed at the boundary as well as the interior nodes to form sufficient solution system. The resulting simultaneous linear system is larger in size than that of the iterative method. But recurrence procedure is not needed. According to Ref. [18], for slow varying permeability relatively few interior nodes are needed to get accurate results. This method can be competitive.

6.3.3.3 Direct Green's Function Scheme

Using the generalized Green's theorem [21], boundary integral equations can be formed for linear, variable coefficient, partial differential equations of any order, so long as the adjoint operators can be found. Greenberg [13] gave the boundary integral equation for the linear, variable coefficient, second-order, partial differential eqution with two independent variables, which has wide engineering applications. The major obstacle for the numerical implementation of the integral equations, however, lies in the fact that the free-space Green's function corresponding to the operator and the set of variable coefficients must be analytically found in order to develop a practical numerical scheme. This, in general, is a difficult mathematical task. We discuss herein a few such Green's functions corresponding to Eq. (6).

The boundary integral equation solving Eq. (6) in two and three spatial dimensions is the following [3, 6]

$$\alpha H = \int \left(KG \frac{\partial H}{\partial n} - KH \frac{\partial G}{\partial n} \right) dS \tag{20}$$

where G is free-space Green's function defined as

$$\nabla \cdot (K \nabla G) = - \begin{Bmatrix} 2\pi \\ 4\pi \end{Bmatrix} \delta(\bar{x}; \bar{x}'). \tag{21}$$

It is clear that Eq. (20) is strictly a boundary equation which does not contain any volume integral. It can be efficiently treated. In fact, due to the similarity between the equation for the homogeneous case, Eq. (8), and the current equation, (20), identical computer codes can be used, except that appropriate kernel functions relating to the set of permeability and direct Green's function should be defined in the function statements of the program.

Analytical determination of Green's function for arbitrary permeability is difficult. In engineering applications, however, permeabilities are usually given as discrete values from field sampling. The resultant can be fitted by some simple functions. In that sense, Green's function needs to be determined only for a few of permeability functions which are suitable for data fitting of the field measurements. We present a few of such results in the following.

Clements [6] derived the two-dimensional Green's function for the one-dimensional, isotropic permeability profile $K(x) = K^0(1 + \beta x)^\eta$, where $K^0 = K(0)$, β and η are parameters of fit. The result is written as a complex variable series

$$K(x')G(x', y'; x, y)$$

$$= -\frac{(1 + \beta x')^{\eta/2}}{(1 + \beta x)^{\eta/2}} \ln r - \frac{(1 + \beta x')^{\eta/2}}{(1 + \beta x)^{\eta/2}} \operatorname{Re} \left\{ \sum_{n=1}^{\infty} \left[\frac{(-1)^n \beta^n (1 + \beta x')^{-n}}{8^n n(n-1)!} \eta(\eta - 2n) \right. \right.$$

$$\cdot \prod_{j=1}^{n-1} (\eta^2 - 4j^2) \cdot \sum_{m=0}^{n-1} \frac{(n-1)!(-1)^m}{m!(n-1-m)!(m+1)} (z' - z)^{n-1-m}$$

$$\cdot ((z' - z)^{m+1} \ln(z' - z) - (-z)^{m+1} \ln(-z) - (m+1)^{-1}(z' - z)^{m+1}$$

$$\left. \left. + (m+1)^{-1}(-z)^{m+1}) \right] \right\}. \tag{22}$$

In the above, $z = x + iy$; $z' = x' + iy'$; x, y are the Cartesian coordinates for the base point; x' and y' are the coordinates for the field point; and Re denotes the real part of the complex expression. The above infinite series truncates into a finite one

Table 1. Green's Functions

Case	K	Permeability variation	Number of parameters	Problem geometry	$K(\bar{x}')G(\bar{x}';\bar{x})$
1	$K^0(1 + \beta x)$	1-D	2	2-D	Eq. 22 with $\eta = 1$
2	$K^0(1 + \beta x)^2$	1-D	2	2-D	$-\dfrac{(1 + \beta x')}{(1 + \beta x)}\ln r$
2a				3-D	$\dfrac{(1 + \beta x')}{(1 + \beta x)}\dfrac{1}{r}$
3	$K^0(1 + \beta x)^{-2}$	1-D	2	2-D	$-\dfrac{(1 + \beta x)}{(1 + \beta x')}\ln r + \dfrac{(1 + \beta x)\beta}{(1 + \beta x')} \cdot$ $\mathrm{Re}\{(z' - z)\ln(z' - z) + z\ln(-z) - z'\}$
4	$K^0(1 + \beta x)^{\eta}$	1-D	3	2-D	Eq. 22
5	$K^0 e^{\beta x}$	1-D	2	2-D	$e^{\beta(x'-x)/2}K_0\left(\dfrac{|\beta|r}{2}\right)$
5a				3-D	$e^{\beta(x'-x)/2}e^{-|\beta|r/2}\dfrac{1}{r}$
6	$K^0 \cos^2 \beta x$	1-D	2	2-D	$-\dfrac{\pi}{2}\dfrac{\cos \beta x'}{\cos \beta x}Y_0(|\beta|r)$
6a				3-D	$\dfrac{\cos \beta x'}{\cos \beta x}\dfrac{\cos \beta r}{r}$
7	$K^0(e^{\beta x} + \eta \sinh \beta x)^2$	1-D	3	2-D	$\dfrac{e^{\beta x'} + \eta \sinh \beta x'}{e^{\beta x} + \eta \sinh \beta x}K_0(|\beta|r)$
7a				3-D	$\dfrac{e^{\beta x'} + \eta \sinh \beta x'}{e^{\beta x} + \eta \sinh \beta x}\dfrac{e^{-|\beta|r}}{r}$
8	$K^0(\cos \beta x + \eta \sin \beta x)^2$	1-D	3	2-D	$-\dfrac{\pi}{2}\dfrac{\cos \beta x' + \eta \sin \beta x'}{\cos \beta x + \eta \sin \beta x}Y_0(|\beta|r)$
8a				3-D	$\dfrac{\cos \beta x' + \eta \sin \beta x'}{\cos \beta x + \eta \sin \beta x}\dfrac{\cos \beta r}{r}$
9	$K^0(c_1 x + c_2 xy + c_3 y + 1)^2$	2-D	4	2-D	$-\dfrac{c_1 x' + c_2 x'y' + c_3 y' + 1}{c_1 x + c_2 xy + c_3 y + 1}\ln r$
9a				3-D	$\dfrac{c_1 x' + c_2 x'y' + c_3 y' + 1}{c_1 x + c_2 xy + c_3 y + 1}\dfrac{1}{r}$
10	$(c_1 e^{\beta x} + c_2 e^{-\beta x})^2 \cdot (c_3 e^{\eta y}$ $+ c_4 e^{-\eta y})^2 = K(x, y)$	2-D	6	2-D	$\dfrac{K^{1/2}(x', y')}{K^{1/2}(x, y)}K_0(|\beta + \eta|r)$
10a				3-D	$\dfrac{K^{1/2}(x', y')}{K^{1/2}(x, y)}\dfrac{e^{-|\beta + \eta|r}}{r}$
11	$K^0(c_1 x + c_2 y + c_3 z + c_4 xy$ $+ c_5 xz + c_6 yz + c_7 xyz + 1)^2$ $= K(x, y, z)$	3-D	8	3-D	$\dfrac{K^{1/2}(x', y', z')}{K^{1/2}(x, y, z)}\dfrac{1}{r}$

if $\eta = \pm 2N$, where N = integer. The special results of $\eta = 2$ and -2 are listed in Table 1. Clements and Rogers [7] also gave similar integral equation and Green's functions for anisotropic permeability.

Using a simple transformation suggested in Ref. [12], Cheng [3] derived closed-form Green's function for a class of permeabilities whose square roots satisfy the Laplace or the Helmholtz equations. The possible functions in this class are virtually limitless. For instance, the real part of any analytic complex function satisfies the Laplace equation. But realistically, only the functions that are suitable for data fitting of field measurements will be used.

In Table 1 we list a few Green's functions in two and three dimensional geometries, with permeability variations in one, two and three dimensions. The following notations are used: $r = |\bar{x} - \bar{x}'|$ for two and three-dimensional geometries; Y_0 and K_0 = Bessel function and modified Bessel function of second kind of order zero, respectively; and K^0 = hydraulic conductivity at the origin. In each case, we also list the number of parameters available for data fitting.

It is of interest to note that the linear, two-parameter profile, case 1 in Table 1, can be approximated by profiles with more than two parameters. For instance, if profile 7 is required to pass through the two end points and the middle point of a linear profile, the error involved in the approximation is no more than a few percent everywhere. Since Green's function of case 7 is in closed form, in contrast to the infinite series of case 1, it may be more advantageous using the approximation than using the exact linear profile.

Details of implementation and application to groundwater flow problems may be found in Ref. [3].

6.4 Heterogeneous Fluid

We discuss herein a few special cases of heterogeneous fluid flows in homogeneous porous media. The fluids may be miscible or immiscible. The mass and heat transfer effects are ignored. In fact, we deal with mainly two cases: a steady state stratified fluid flow and the transient interface of two immiscible fluids.

6.4.1 Steady-state Miscible Fluid Flow

The miscible fluid flows without mass and heat transfer are governed by Eqs. (1), (3), (4) and (5). Efforts have been made to reduce those equations to Poisson type equations. For instance, under constant viscosity Eqs. (1) and (5) can be combined to yield the following equations given by Knudsen [14]

$$\nabla^2 p = -g\frac{\partial p}{\partial z} \tag{23}$$

and by de Josselin de Jong [10, 11]

$$\nabla^2 \theta = \frac{kg}{\mu}\nabla \cdot (z\nabla\rho) \tag{24}$$

where $\theta = (k/\mu)(p + \rho gz)$. Except for the special cases in which the density distribution is known a priori, Eqs. (23) and (24) must be coupled with Eq. (3) for solution. This may pose difficulty for boundary element treatment. In a special steady state case, however, a strict boundary method results and the numerical implementation should be very efficient.

Following Yih [31], under steady state condition the viscosity can be incorporated into the velocity thus yielding an "associated flow" governed by the following equations

$$\frac{\mu_0}{k}q' = -\nabla p - \rho g \nabla z \tag{25}$$

$$q' \cdot \nabla \rho = 0 \tag{26}$$

$$\nabla \cdot q' = 0 \tag{27}$$

where $q' = \mu q/\mu_0$ = associated velocity; and μ_0 = a reference viscosity. It is noted that Eqs. (25) and (27) are precisely the equations governing the flow of constant viscosity, with the velocity replaced by the associated velocity. Therefore, the effect of the viscosity is to reduce the local velocity by a factor of μ_0/μ.

In two-dimensional geometry, Yih [31] introduced the stream function based on the associated flow

$$q'_x = \frac{\partial \psi'}{\partial z}, \quad q'_z = -\frac{\partial \psi'}{\partial x} \tag{28}$$

Substituting the above into (25), cross differentiating, and combining with Eq. (26) yield

$$\nabla^2 \psi' = \frac{kg}{\mu_0} \frac{d\rho}{d\psi'} \frac{\partial \psi'}{\partial x} \tag{29}$$

where the density is a known function of ψ' and is given as part of the boundary condition. Equation (29) is in general nonlinear. But as pointed out in Ref. [31], if the flow originates far upstream as a horizontal flow with linear density stratification $\rho = \rho_0 - (\rho_0 - \rho_1)z/d$, where ρ_0 is the density at the lower boundary ($z = 0$) and ρ_1 is the density at the upper boundary ($z = d$), Eq. (29) is reduced to the linear elliptic equation

$$\nabla^2 \psi' + \beta \frac{\partial \psi'}{\partial x} = 0 \tag{30}$$

where $\beta = kg(\rho_0 - \rho_1)/\mu_0 U'd$, and U' = uniform flow velocity of the associated flow far upstream.

Equation (28) can be subjected to a BEM treatment. Applying Green's reciprocal theorem we obtain the following

$$\alpha\psi' = \int \left[\phi \frac{\partial \psi'}{\partial n} - \psi' \left(\frac{\partial \psi'}{\partial n} - \beta\phi n_x \right) \right] dS \tag{31}$$

where n_x is the x-component of the outward normal, and ϕ is Green's function of the adjoint operator

$$\nabla^2 \phi - \beta \frac{\partial \phi}{\partial x} = -2\pi \delta(\bar{x}; \bar{x}').$$ (32)

Solution of Eq. (32) has been given [4]

$$\phi(\bar{x}; \bar{x}') = e^{\beta(x-x')/2} K_0(\beta r/2).$$ (33)

6.4.2 Transient Interface of Immiscible Fluid

6.4.2.1 Multizone Approach

For immiscible fluid, the viscosity and the density remain constant within each fluid zone. Further assuming constant permeability, the Laplace equation is applicable in each fluid domain. The multizone BEM approach as described in Sect. 6.3.2.1 can be used. In other words, the BEM is applied once for each zone. The resultants are then coupled and condensed using the interfacial conditions of flux and pressure continuities. After the simultaneous system is solved, the flux at the interface can be used to move the interface quasi-statically to a new location. The new interface then defines the geometry for the next level of solution. By time stepping, the successive locations of the interface can be recorded. The above procedure has been successfully adopted for the prediction of the movement of sea-fresh water interface in coastal aquifers [20, 29].

6.4.2.2 Single Potential Approach

Since the pressure is continuous across an immiscible interface, a scheme analogous to that in Sect. 6.3.2.2 can be developed to reduce the size of the solution matrix.

Through elimination of q and some manipulation, Eqs. (1) and (5) can be combined to give the following

$$\nabla^2 p = \nabla \ln \mu \cdot \nabla p + \rho g \frac{\partial \ln \mu}{\partial z} - g \frac{\partial \rho}{\partial z}$$ (34)

where ρ and μ are functions of space. Treating Eq. (34) as a Poisson equation, a boundary integral equation may be written with the right hand side of Eq. (34) appearing in a volume integral. The volume integral is null everywhere due to constant density and viscosity, except at the interface where finite jumps take place. Following the limiting processes described in Sect. 6.3.2.2 we arrive at the following integral equation

$$\alpha p = \int \left(G^* \frac{\partial p}{\partial n} - p \frac{\partial G^*}{\partial n} \right) dS + \int G^* \ln \frac{\mu_2}{\mu_1} \frac{\partial p}{\partial n} dS_i$$

$$+ \int G^* g n_z \left(\frac{\rho_1 + \rho_2}{2} \ln \frac{\mu_2}{\mu_1} + \rho_1 - \rho_2 \right) dS_i.$$ (35)

In the above, the subscripts 1 and 2 denote parameters associated to the respective zones, and n_z is the z-component of the interfacial normal.

It is noted that in Eq. (35) the integration is performed along the external boundary S and the interface S_i. There is no volume integral involved. Equation (35) is applied when base points are located at external boundary. For base points at the interior boundary, we differentiate Eq. (35) with respect to the interfacial normal. The resulting system eliminate the need to formulate the pressure at the interface nodes and only the normal gradient is solved there. The final matrix is $(M + N) \times (M + N)$, where M and N are, respectively, the number of exterior and interior nodes.

It is of particular interest to note that in the application to saltwater encroachment in coastal aquifers, the viscosity variation between the salt and the fresh water is usually negligible. In this case, Eq. (35) is further reduced

$$\alpha p = \int \left(G^* \frac{\partial p}{\partial n} - p \frac{\partial G^*}{\partial n} \right) dS + \int G^* g n_z (\rho_1 - \rho_2) \, dS_i. \tag{36}$$

It is clear from the above that there is no unknown involved at the interface. Although numerical integration is needed along the interface, the resultant matrix is $M \times M$ only. We may also view the second integral of Eq. (36) as that the jump in density has been simultated by distributing sources along the interface.

6.5 Embedded Heterogeneity

In this section we consider the inclusion of material in the form of thin laminae in an otherwise homogeneous porous medium. The embedded material may be an impermeable sheetpile used for the reduction of seepage flow underneath an earth dam [24], a semipermeable clay lense dividing aquifers [8, 22, 27] or a finite conductivity fracture surface used for enhancing oil recovery [5, 25]. Those heterogeneities are usually formulated as surfaces across which discontinuities in potential and/or flux take place.

Because the heterogeneities are isolated, one is tempted to enclose and to exclude them for the solution domain. This yields a homogeneous, multiply-connected region. The BEM is directly applied with nodes distributed on the exterior as well as the interior boundaries. In the limiting case we should shrink the contours to coincide with the discontinuity surfaces. As discussed in Ref. [17], the resulting BEM matrix is singular because nodes on the two sides of the contour occupy the same location. Artificially separating the nodes on the two sides by a small distance does not solve the problem. In the numerical experiment of varying the gap thickness divergent results which are strongly dependent on the actual thickness separating the nodes are observed. It is apparent that more careful treatment is required. We discuss two such measures in the following.

6.5.1 Multizone Approach

To create a multizone situation, the solution domain is subdivided using not only the embedded surfaces but also some arbitrary extensions. Although there is no discontinuity along the extensions, they are formally treated as interzonal boundaries in the multizone approach.

Lafe et al. [17] used this approach to solve problems of seepage flow around sheetpile. The domain is subdivided along the cutoff wall and its extension. The no flux condition is enforced on both sides of the wall and the two potentials, one on each side, are solved. Along the extension, both the potential and the flux are continuous. One potential and one flux are then solved at each of the nodes.

Shapiro and Andersson [26] investigated the steady state porous media flow in fractured rock. The fracture is modeled as infinitesimally thick with finite hydraulic conductivity along the direction of the crack. The potential is continuous across the gap but the normal flux has finite jump due to the flow along the fracture. From continuity and one-dimensional viscous flow considerations, the following equation governs the fracture flow:

$$\frac{\partial}{\partial s} \kappa \frac{\partial H}{\partial s} = K \left(\frac{\partial H}{\partial n} \right)_l + K \left(\frac{\partial H}{\partial n} \right)_r. \tag{37}$$

In the above s is the measure of length along the fracture; $\kappa =$ the fracture conductance; and the subscripts l and r denote the left and the right side of the fracture, respectively. Using the above as the coupling condition, the multizone BEM is again applicable. Details of the implementation may be found in the original reference.

6.5.2 Singularity Distribution

Distribution of singularity of physical or fictitious strength has long been used as a way to a simulate surfaces with discontinuities. In elasticity, for example, it is well known that distributed dislocations simulate fracture surfaces [2]. In potential flows, doublets and vortices have been used to create jumps in properties such as density [10] and permeability [28].

As discussed in Ref. [28], discontinuity along a line can be excluded from a domain by performing a path integral around it. If jumps of both the potential and the flux are observed across the line, the boundary integral equation can be formally written as

$$\alpha H = \int \left(G^* \frac{\partial H}{\partial n} - H \frac{\partial G^*}{\partial n} \right) dS$$

$$+ \int \left(G^* \frac{\partial H^+}{\partial n^+} - H^+ \frac{\partial G^*}{\partial n^+} \right) dS^+ + \int \left(G^* \frac{\partial H^-}{\partial n^-} - H^- \frac{\partial G^*}{\partial n^-} \right) dS^-$$

$$= \int \left(G^* \frac{\partial H}{\partial n} - H \frac{\partial G^*}{\partial n} \right) dS + \int G^* \left(\frac{\partial H^+}{\partial n^+} + \frac{\partial H^-}{\partial n^-} \right) dS^+$$

$$- \int (H^+ - H^-) \frac{\partial G^*}{\partial n^+} dS^+ \tag{38}$$

where S is the external boundary; S^+ and S^- are the contours on the two sides of the line element; and the superscripts $+$ and $-$ denote quantities defined on the contours S^+ and S^-, respectively.

The first integral on the right side of Eq. (38) is identical to the boundary integral equation for homogeneous material. The second and the third integrals suggest additional contributions from distributed sources and doublets, respectively, along the line element. The strengths of the distributions are physical quantities, namely the jumps of the potential and its normal derivative. This makes the above equation especially attractive, since no fictitious quantities are involved.

We demonstrate herein the BEM formulation for flow in fractured porous media. Along the fracture, the potential is continuous and Eq. (38) is reduced to

$$\alpha H = \int \left(G^* \frac{\partial H}{\partial n} - H \frac{\partial G^*}{\partial n} \right) dS + \int G^* \left(\frac{\partial H^+}{\partial n^+} + \frac{\partial H^-}{\partial n^-} \right) dS^+ \tag{39}$$

Substitution of (37) to the above yields

$$\alpha H = \int \left(G^* \frac{\partial H}{\partial n} - H \frac{\partial G^*}{\partial n} \right) dS + \int \frac{G^*}{K} \frac{\partial}{\partial s} \kappa \frac{\partial H}{\partial s} dS^+ \tag{40}$$

Equation (40) can be readily subjected to the regular BEM treatment of discretization, interpolation using shape functions, and numerical integration. Base points are distributed both on the exterior boundary, S, and on the fracture surface, S^+. Since only discrete values of H are formulated on the fracture surface, the resulting system is $(M + N) \times (M + N)$ and is sufficient for solution. In this scheme we have eliminated the need for nodes on arbitrary extensions and halved the unknown on the fracture surface.

6.6 Concluding Remarks

In this article we reviewed several ideas for dealing with material heterogeneity by the Boundary Element Method. Although the discussion above was limited to flow through porous media, some of the ideas can be readily extended to elasticity, as well as other type of governing equations.

We demonstrated in the above that in many cases heterogeneity does not pose difficulty for the BEM and can be efficiently treated. We acknowledge, however, that the conditions under which some of the proposed methods apply are still somewhat restrictive. More work in this area is needed to further improve the competitiveness of the BEM.

References

1 Bettess, J.A. (1983), "Economical Solution Technique for Boundary Integral Matrices," Int. J. Num. Meth. Engrg., Vol. 19, pp. 1073–1077.

2 Bilby, B.A. (1960), "Continuous Distributions of Dislocations," Progress in Solid Mechanics, Vol. 1, eds., I.N. Sneddon and R. Hill, North-Holland, London, pp. 329–398.

3 Cheng, A.H-D. (1984), "Darcy's Flow with Variable Permeability: A Boundary Integral Solution," Water Resources Research, Vol. 20, No. 7, pp. 980–984.

4 Cheng, A.H-D. and Liggett, J.A. (1984), "Boundary Integral Equation Method for Linear Porous-Elasticity with Applications to Fracture Propagation," Int. J. Num. Meth. Engrg., Vol. 20, No. 2, pp. 279–296.

5 Cinco-Lay, H., Samaniego-V., F. and Dominguez-A., N. (1978), "Transient Pressure Behavior of a Well with a Finite-Conductivity Vertical Fracture," Society of Petroleum Engineers Journal, August, pp. 253–264.

6 Clements, D.L. (1980), "A Boundary Integral Equation Method for the Numerical Solution of a Second Order Elliptic Equation with Variable Coefficients," J. Austral. Math. Soc., Vol. 22, Ser. B, pp. 218–228.

7 Clements, D.L. and Rogers, C. (1983), "A Boundary Integral Equation for the Solution of a Class of Problems in Anisotropic Inhomogeneous Thermostatics and Elastostatics," Quarterly of Applied Mathematics, Vol. 41, No. 1, pp. 99–105.

8 Collins, M.A. and Gelhar, L.Y. (1971), "Seawater Intrusion in Layered Aquifers," Water Resources Research, Vol. 7, No. 4, pp. 971–979.

9 Crotty, J.M. (1982), "A Block Equation Solver for Large Unsymmetric Matrices Arising in the Boundary Integral Equation Method," Int. J. Num. Meth. Engrg., Vol. 18, pp. 997–1017.

10 de Josselin de Jong, G. (1960), "Singularity Distributions for the Analysis of Multiple Fluid Flow Through Porous Media," J. Geophys. Res., Vol. 65, pp. 3739–3758.

11 de Josselin de Jong, G. (1969), "Generating Functions in the Theory of Flow Through Porous Media," in Flow Through Porous Media, ed. R.J.M. De Wiest, Academic Press, New York.

12 Georghitza, St.I. (1969), "On the Plane Steady Flow of Water Through Inhomogeneous Porous Media," the First Symposium on the Fundamentals of Transport Phenomena in Porous Media, Int. Assoc. Hydraul. Res., Haifa, Israel.

13 Greenberg, M.D. (1971), Application of Green's Functions in Science and Engineering, Prentice-Hall.

14 Knudsen, W.C. (1962), "Equations of Fluid Flow Through Porous Media-Incompressible Fluid of Varying Density," J. Geophys. Res., Vol. 67, pp. 733–737.

15 Lachat, J.C. and Watson, J.O. (1977), "Progress in the Use of Boundary Integral Equations, Illustrated by Examples," Computer Meth. Appl. Mech. and Engrg., Vol. 11, pp. 1753–1768.

16 Lafe, O.E., Liggett, J.A. and Liu, P.L-F. (1981), "BIEM Solutions to Combinations of Leaky, Layered, Confined, Unconfined, Nonisotropic Aquifers," Water Resources Research, Vol. 17, No. 5, pp. 1431–1444.

17 Lafe, O.E., Montes, J.S., Cheng, A.H-D., Liggett, J.A. and Liu, P.L-F. (1980), "Singularities in Darcy Flow Through Porous Media," Journal of the Hydraulics Division, ASCE, Vol. 106, No. HY6, pp. 977–997.

18 Lennon, G.P. (1984), "Boundary Element Analysis of Flow in Hetrogeneous Porous Media," Proc. ASCE/HYD Specialty Conf., Coeur d'Alene, Idaho.

19 Liggett, J.A. and Liu, P.L-F. (1983), The Boundary Integral Equation Method for Porous Media Flow, George Allen and Unwin, London.

20 Liu, P.L-F., Cheng, A.H-D., Liggett, J.A. and Lee, J.H. (1981), "Boundary Integral Equation Solutions to Moving Interface Between Two Fluids in Porous Media," Water Resources Research, Vol. 17, No. 5, pp. 1445–1452.

21 Morse, P. and Feshbach, H. (1953), Methods of Theoretical Physics, Parts I and II, McGraw-Hill, New York.

22 Mualem, Y. and Bear, J. (1974), "The Shape of the Interface in Steady Flow in a Stratified Aquifer," Water Resources Research, Vol. 10, No. 6, pp. 1207–1215.

23 Mukherjee, S. (1982), Boundary Element Methods in Creep and Fracture, Applied Science Publishers, London.

24 Polubarinova-Kochina, P.Ya. (1962), Theory of Ground Water Movement, translated by J.M.R. De Wiest, Princeton University Press, Princeton, New Jersey.

25 Prats, M. (1961), "Effect of Vertical Fractures on Reservoir Behavior – Incompressible Fluid Case," Society of Petroleum Engineers Journal, Trans., AIME, Vol. 222, pp. 105–118.

26 Shapiro, A.M. and Andersson, J. (1983), "Steady State Fluid Response in Fractured Rock: A Boundary Element Solution for a Coupled Discrete Fracture Continuum Model," Water Resources Research, Vol. 19, No. 4, pp. 959–969.

27 Strack, O.D.L. (1981), "Flow in Aquifers with Clay Laminae, 1. The Comprehensive Potential," Water Resources Research, Vol. 17, No. 4, pp. 985–992.

28 Strack, O.D.L. and Haitjema, H.M. (1981), "Modeling Double Aquifer Flow Using a Comprehensive Potential and Distributed Singularities, 2. Solution for Inhomogeneous Permeabilities," Water Resources Research, Vol. 17, No. 5, pp. 1551–1560.

29 Taigbenu, A.E., Liggett, J.A. and Cheng, A.H-D. (1984), "Boundary Integral Solution to Seawater
    Intrusion into Coastal Aquifers," Water Resources Research, Vol. 20, No. 8, pp. 1150–1158.
30 Wei, L-Y. and Liggett, J.A. (1982), "Zoned Boundary Elements – An Economical Calculation," Proc.
    Int. Conf. on Finite Element Methods, Shanghai, China.
31 Yih, C-S. (1980), Stratified Flows, Academic Press, New York.
32 Zienkiewicz, O.C., Kelly, D.W. and Bettes, P. (1977), "The Coupling of the Finite Element Method
    and Boundary Solution Procedures," Int. J. Num. Meth. Engrg., Vol. 11, pp. 355–375.

Chapter 7

Time Dependent Ground Flow Analysis

by K. Mizumura

Abstract

Boundary element method is employed to analyze the effects of spacial distribution
of soil infiltration properties and rainfall rate on the hydrologic performance of
catchment area. For simplicity these are modeled by recharge problem to the
groundwater flow. The simulation model also can determine hydrograph bias due
to variability of soil infiltration properties and rainfall rate in space. Further, the
hydrograph bias due to movement and time variation of recharge source is also
studied.

7.1 Introduction

In the rainfall-runoff system a hydrograph has four elements: (1) direct surface
runoff, (2) subsurface stream flow, (3) groundwater, and (4) channel precipitation.
Since the infiltrated component of rainfall is considered to be recharge source to
groundwater flow, the problem of groundwater flow through an unconfined aquifer
receiving vertical recharge is studied as the part of the rainfall-runoff system. The
first application of the boundary element method (Brebbia 1978, abbreviated as
BEM hereafter) to the recharge problem (Hunt 1971) was done by Liggett (1977)
and Liggett and Liu (1983). In this study, the computational method of groundwater
problem by BEM is applied to hydrology and the effect of rainfall position on
hydrograph is investigated (Mizmura 1985). This is related with the following
problems: (1) fast-moving thunder storm and (2) snowmelt runoff. Herein, the
correspondence between the different types of recharge source and the discharge
hydrograph is analyzed by BEM. That is, the effect of spacial and time distribution
of rainfall, spacial distribution of soil infiltration properties (Smith et al. 1979),
rainfall-region movement are studied. To obtain qualitative understanding on the
groundwater flow under the condition of space-varying, time-varying and moving
recharge source in the watershed, a simplified model is numerically and experi-
mentally tested.

7.2 Mathematical Formulation

The motion of a fluid in porous media is considered in the simple region as shown
in Fig. 1. The watershed including the recharge area is assumed to be a two-
dimensional strip of finite width. The velocity potential ϕ is defined for this case as

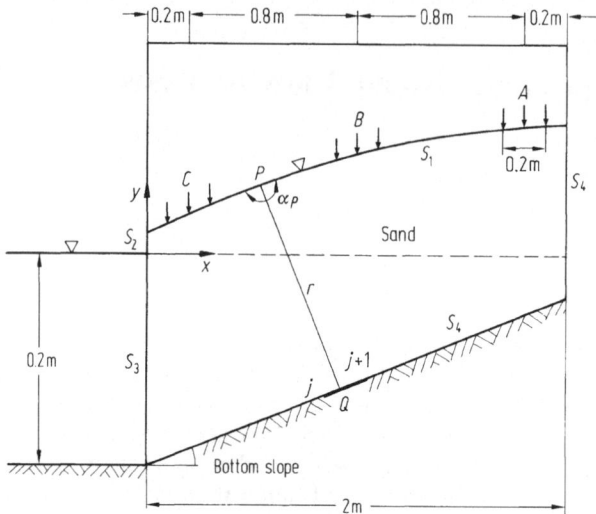

Fig. 1. Definition sketch

$$\phi = y + \frac{p}{w} \tag{1}$$

in which y = the vertical distance from the initial free surface; p = fluid pressure; and w = the fluid specific weight. If the flow in the porous media is described by Darcy's law (Muskat 1937), the governing equation becomes the Laplace equation as follows:

$$\nabla^2 \phi(x, y, t) = 0 \tag{2}$$

in which $\nabla^2 = \partial^2/\partial x^2 + \partial^2/\partial y^2$ and x, y = Cartesian coordinates as shown in Fig. 1. To use BEM, let us consider the following boundary conditions as represented in Fig. 1:

$$p = 0 \ (\phi = \eta) \quad \text{on } S_1 \tag{3}$$

$$p = 0 \ (\phi = y) \quad \text{on } S_2 \tag{4}$$

$$\phi = 0 \quad\quad\quad \text{on } S_3 \tag{5}$$

$$\frac{\partial \phi}{\partial n} = 0 \quad\quad \text{on } S_4 \tag{6}$$

in which $\eta(x, t)$ = the elevation of the free surface above the initial free surface and n = a unit outward normal. Equations (3) and (4) represent that the pressure is atmospheric along the free surface; Eq. (5) prescribes the hydrostatic pressure; and Eq. (6) shows that the aquifer bottom and the side wall are impermeable. For the start of the computation, initial values of ϕ and η should be given. We assume that the fluid is at rest at $t = 0$. Therefore, we have

$$\phi = \eta = 0 \quad \text{at } t = 0. \tag{7}$$

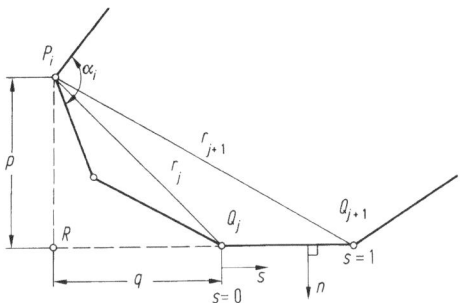

Fig. 2. Definition sketch on each element

By using the fundamental solution of the two-dimensional Laplace equation and Eq. (2), the two-dimensional Green's formula is written as follows:

$$\alpha_P \phi_P + \int_S \phi \frac{\partial}{\partial n}\left(\ln\frac{1}{r}\right) ds - \int_S \frac{\partial \phi}{\partial n} \ln\frac{1}{r} ds = 0 \tag{8}$$

in which P = the nodal point on the boundary; α_P = the angle between two tangents at P; ϕ_P = the value of the velocity potential at P; r = the distance between the control point P and the observation point Q on a line element; and s = the local coordinate measured on the line element as shown in Fig. 2. The substitution of the boundary conditions (3), (4), (5), and (6) into Eq. (8) yields the following integral equation:

$$\alpha_P \phi_P + \int_{S_1+S_2+S_4} \phi \frac{\partial}{\partial n}\left(\ln\frac{1}{r}\right) ds - \int_{S_1+S_2+S_3} \frac{\partial \phi}{\partial n} \ln\frac{1}{r} ds = 0. \tag{9}$$

7.3 Free Surface Calculations

In addition to the above boundary conditions, the following equation must be satisfied on the free surface:

$$\theta_e \frac{\partial \eta}{\partial t} = -\frac{k}{n_y} \frac{\partial \phi}{\partial n} + R \tag{10}$$

in which θ_e = the effective porosity at the free surface; k = the coefficient of permeability; n_y = the direction cosine of the normal with respect to the y-axis; and R = the rate of recharge.

7.4 Discretization

To discretize Eqs. (9) and (10), the functions ϕ and η are assumed to be linear functions of the local coordinate s in a linear element. The line element represents a line segment produced by discretization. By using the above assumption,

Eq. (9) becomes

$$\alpha_i \phi_i + \sum_{S_1+S_2+S_4} \int_0^l N^T \frac{\partial}{\partial n}\left(\ln \frac{1}{r}\right) ds \cdot \boldsymbol{\phi} - \sum_{S_1+S_2+S_3} \int_0^l N^T \ln \frac{1}{r} ds \cdot \boldsymbol{f} = 0$$

$$\text{for } i = 1, 2, \ldots, M \qquad (11)$$

in which i = the subscript of the control point; M = total number of nodal points; l = the length of the line element; $N^T = [(l - s)/l, s/l]$; $\boldsymbol{\phi}^T = [\phi_j, \phi_{j+1}]$; and $\boldsymbol{f}^T = [\partial\phi/\partial n|_j, \partial\phi/\partial n|_{j+1}]$. The free surface condition of Eq. (10) is written using the finite differences with respect to time as follows:

$$\phi_j^{K+1} = \phi_j^K - \frac{k\Delta t}{2n_y\theta_e}\left\{\left(\frac{\partial\phi}{\partial n}\right)_j^{K+1} + \left(\frac{\partial\phi}{\partial n}\right)_j^K\right\} + \frac{R_j^{K+1} + R_j^K}{2\theta_e}\Delta t \qquad (12)$$

in which K = the superscript to represent the time level; and Δt = the time increment. The substitution of Eq. (12) into Eq. (11) gives the following equation:

$$\alpha_i \phi_i^{K+1} + \sum_{S_1} A^T\left\{\boldsymbol{\phi}^K - \frac{k\Delta t}{2n_y\theta_e}\boldsymbol{f}^T + \frac{R^{K+1} + R^K}{2\theta_e}\Delta t\right\}$$

$$- \sum_{S_1}\left\{B^T + \frac{k\Delta t}{2n_y\theta_e}A^T\right\}\boldsymbol{f}^{K+1} - \sum_{S_2+S_3} B^T\boldsymbol{f}^{K+1} + \sum_{S_2+S_4} A^T\boldsymbol{\phi}^{K+1} = 0 \qquad (13)$$

in which $A^T = \int_0^l N^T \frac{\partial}{\partial n}\left(\ln \frac{1}{r}\right) ds$; $B^T = \int_0^l N^T \ln \frac{1}{r} ds$; and $R^T = [R_j, R_{j+1}]$. The above integrals A^T and B^T can be calculated analytically (Nakayama, et al., 1981). They are

$$A^T = \frac{1}{l}[A_L - (q + l)A_T, -A_L + qA_T] \qquad (14a)$$

$$A_L = p \ln(r_{j+1}/r_j) \qquad (14b)$$

$$A_T = \tan^{-1}\left[\frac{pl}{p^2 + q(q + l)}\right] \qquad (14c)$$

$$B^T = \frac{1}{l}[B_L - (q + l)B_T, -B_L + qB_T] \qquad (15a)$$

$$B_L = \tfrac{1}{4}\{r_{j+1}^2(2 \ln r_{j+1} - 1) - r_j^2(2 \ln r_j - 1)\} \qquad (15b)$$

$$B_T = (q + l)\ln r_{j+1} - q \ln r_j - l + pA_T \qquad (15c)$$

in which r_j and r_{j+1} = distances between P_i and Q_j and between P_i and Q_{j+1}, respectively as shown in Fig. 2. p and q are distances between R and P_i, and between R and Q_j, respectively, where R denotes a foot of a perpendicularly drawn from P_i to the extension of the element Q_jQ_{j+1}. When P_i is in the opposite direction to the normal vector n drawn outwardly on the element, p is a negative value. q takes a positive value when R is in the negative side of the local coordinate s. By solving Eq. (13), $\boldsymbol{f}^{K+1}$ on the boundaries S_1, S_2, and S_3 and ϕ^{K+1} on the boundary S_4 are obtained. After solving that, ϕ^{K+1} on the boundary S_1 are calculated by Eq. (12).

7.5 Numerical Results

For numerical calculation the time increment Δt and the coefficient of permeability k are selected 0.5 sec and 0.0001 m/sec, respectively. Figure 3 represents the hydrograph of the runoff discharge calculated at the downstream boundaries S_2 and S_3 for given recharge source at different positions A, B, and C as designated in Fig. 1. This type of recharge source is defined as the fundamental case. As the distance between the position of the recharge source and the boundaries S_2 and S_3 increases, the time to peak discharge increases and the shape of the discharge hydrograph becomes smooth and flat as expected. In the figure the discharge hydrograph of the uniformly distributed recharge source over the whole free surface is shown when the total recharge rate is the same as that located on each position. This is designated by "Averaged" in the figures of this paper. The discharge hydrograph is remarkably influenced by the position of the recharge source near the downstream boundaries. Figure 4 shows discharge hydrographs, which are calculated on the downstream boundaries, when the recharge intensity is a half and the recharge duration is twice

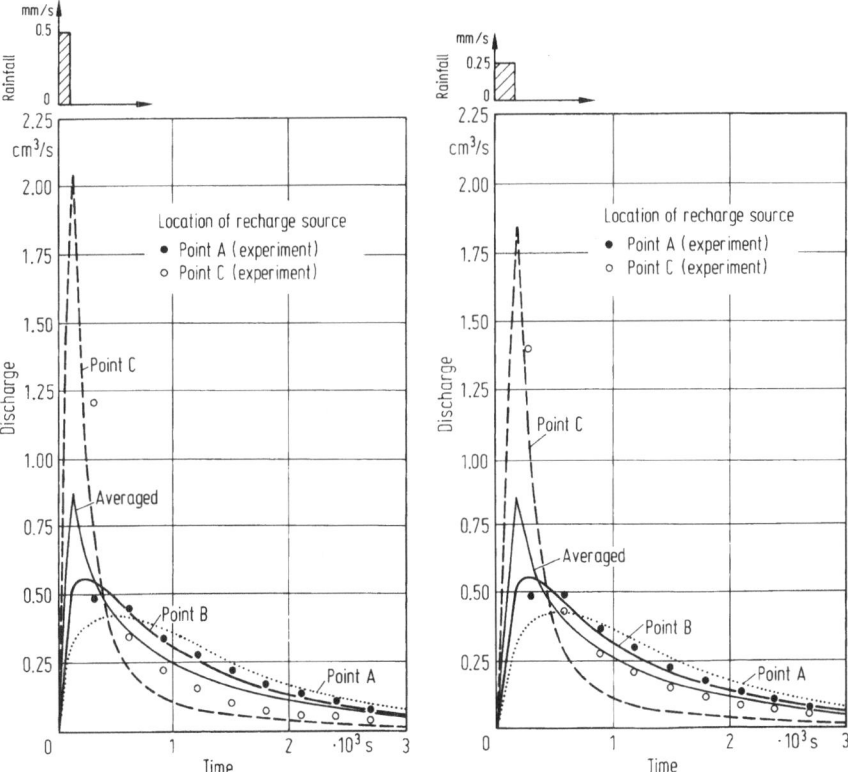

Fig. 3. Discharge hydrograph (including experimental data)

Fig. 4. Discharge hydrograph (including experimental data)

in the fundamental recharge type. The total quantity of recharge rate is the same. The peak discharge decreases and the time to the peak discharge becomes longer when the recharge source is located on Point C. When the recharge source is on point A or B, the peak discharges are almost the same as that in the fundamental case and the time to the peak discharge increases. The characteristics of the discharge hydrograph is almost the same as that in the previous case when the recharge source is uniformly distributed on the whole free surface boundary and the total quantity of the recharge is the same. Figure 5 gives hydrograph in the case when the rate of the recharge intensity becomes twice and the duration is a half in the fundamental case. The change of the recharge intensity corresponds to the discharge hydrograph at Point C sensitively. The previously discussed characteristics of the discharge hydrograph becomes clearer when the hydrographs of the next case are plotted in Fig. 6. In this case the intensity of the recharge rate becomes four times

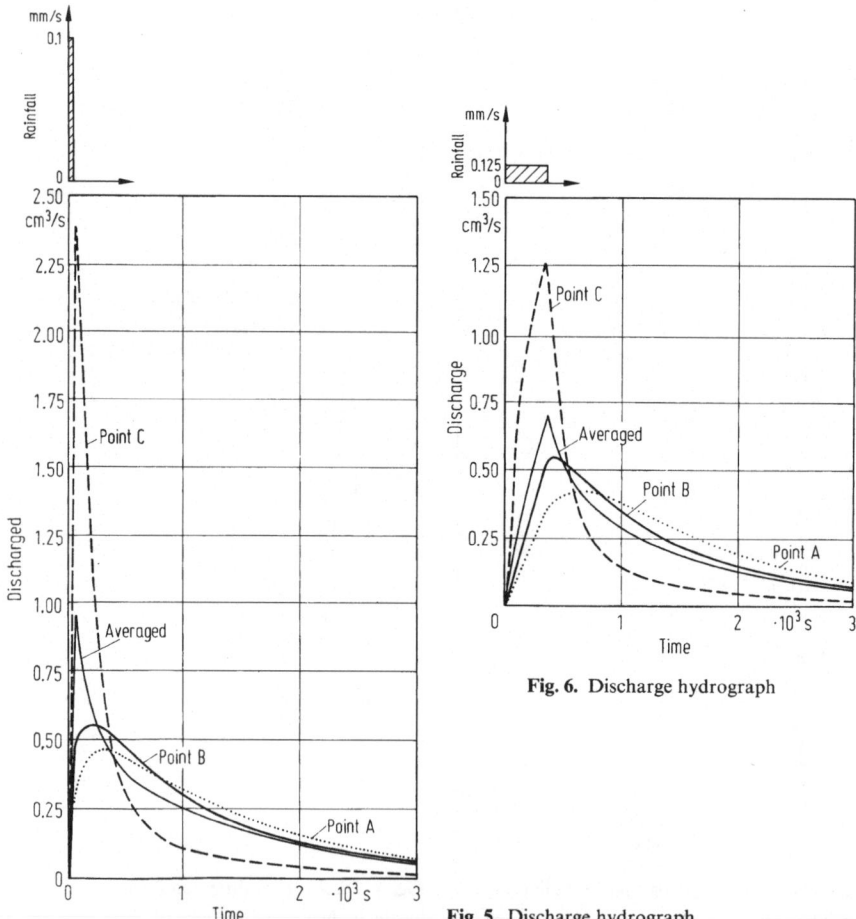

Fig. 6. Discharge hydrograph

Fig. 5. Discharge hydrograph

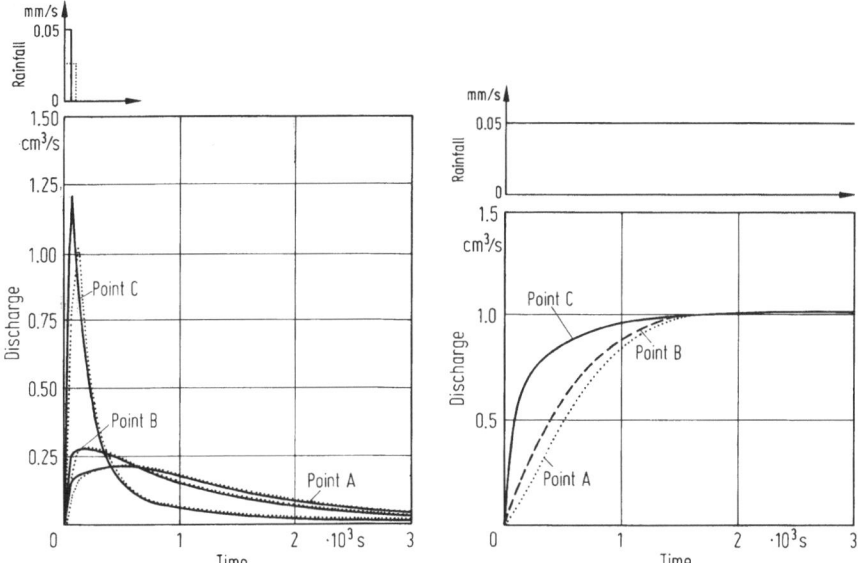

Fig. 7. Comparison of discharge hydrograph   **Fig. 8.** Comparison of rising rimb

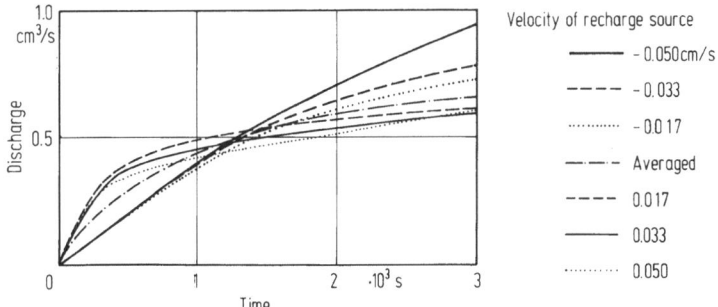

Fig. 9. Effect of moving recharge source on discharge hydrograph

and the duration one fourth that in the fundamental case. Figure 7 gives the comparison of the hydrographs for two kinds of different recharge source. The discrepancy of the discharge hydrographs is remarkable when the recharge source is located on Point C. The difference of the discharge hydrographs are specially found along the rising limb when the recharge sources are even located on the other points A and B. Figure 8 shows the rising limb of the hydrographs when the continuous recharge rate in time are given on three positions A, B, and C. The effect of the location for given recharge source is clearly found. Figure 9 represents the rising limbs of the hydrograph when the recharge source is moving with different velocities in two directions. In the figure the "positive" velocity of the recharge

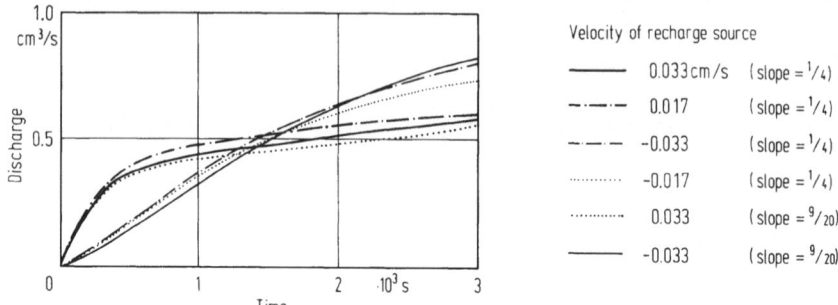

Fig. 10. Effect of moving recharge source on discharge hydrograph

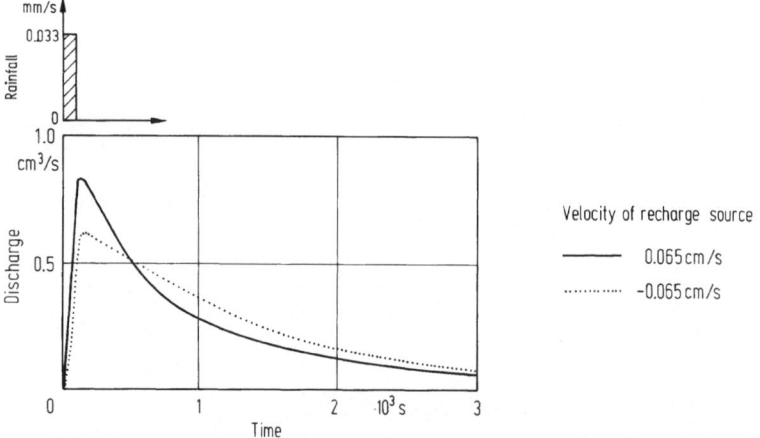

Fig. 11. Effect of moving recharge source on discharge hydrograph

source shows that the recharge source moves from Point C to A and the "negative" one gives the opposite movement of the recharge source. When the recharge source moves in the positive direction, the discharge suddenly increases and then is rising at a decreasing rate. But on the other hand, when the recharge source moves in the negative direction, the discharge increases almost constantly. Figure 10 shows the influence of the different bottom slopes under the same conditions of the recharge source. The existence of the bottom slope amplifies the effect of the movement. Figure 11 gives the deformation of the discharge hydrograph due to the movement of the recharge source. The slope of the discharge hydrograph becomes sharp when the recharge source moves in the positive direction. On the other hand, it becomes smooth for the negative movement of the recharge source. Figure 12 represents the discharge hydrographs when the time-varying recharge rate is given on different positions A, B, and C. As the distance between the position of the recharge source and the downstream boundaries increases, the time variation of the recharge source disappears.

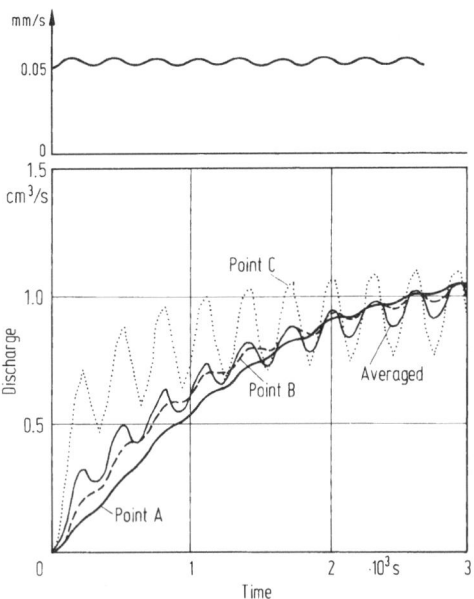

Fig. 12. Effect of time-varying recharge source on discharge hydrograph

7.6 Experimental Investigation

To verify the numerical result obtained by BEM, the experiment was conducted by using a water tank for porous media flow as shown in Fig. 13. The dimensions are as follows: the height, the width, and the length are 1.2 m, 0.2 m, and 2 m, respectively. Drainage hole and orifice are fitted in the water tank to drain water in condition of keeping constant water level. Between sand and subtank with orifice (Fig. 13), mesh is fixed in order that sand does not move out. The flow of constant discharge is spinkled over a piece of sponge (2 cm thickness × 20 cm length × 20 cm width) on the sand surface as rainfall and thus the water infiltrates into the sand through

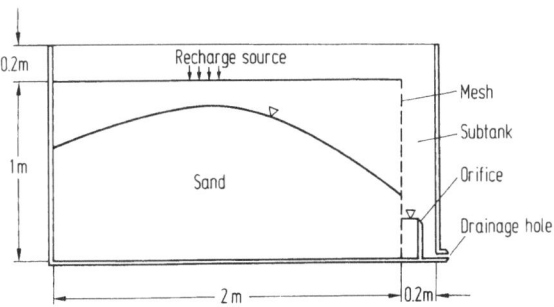

Fig. 13. Experimental apparatus

the sponge. The mean diameter of sand is nearly 0.18 mm. The discharge from the model watershed is measured by taking water at the drainage hole and measuring the weight of the discharged water. The results are plotted in Figs. 3 and 4. The movement of water flow between the free surface and the sand surface is neglected in thus experiment. That is, the time when fluid moves from the sand surface to the free surface of water is assumed to be zero in the computation.

7.7 Summary and Conclusions

Through the study of the groundwater flow analysis under the space-varying, the time-varying, and the moving recharge sources, the following conclusions are obtained:

1) For the given watershed model the location of recharge source influences the discharge hydrograph remarkably. The bias of spacial distribution of rainfall or soil infiltration properties gives the deflection of the discharge hydrograph.
2) Although the phenomena of fluid motion between the sand surface and the free surface of water is neglected, the comparison between the numerical and the experimental result shows satifactory agreement.
3) The movement of the recharge source conspicuously influences the shape of the discharge hydrograph. For large bottom slope the dependence is remarkable.
4) When the location of the recharge source is near the downstream boundary where the discharge is calculated, the time variation of the discharge hydrograph is very sensitive to that of the recharge source. On the other hand, the time dependence of the recharge source on the distant position from the downstream boundary rapidly recedes.

References

Bear, J. 1972. Dynamics of fluids in porous media. New York, American Elsevier.

Brebbia, C.A. 1978. The boundary element method for engineers. Pentech Press Limited.

Hunt, B.W. 1971. Vertical recharge of unconfined aquifer. J. of the HY Div., ASCE, Vol. 97, No. HY7: 1017–1030.

Liggett, J.A. 1977. Locations of free surface in porous media. J. of the HY Div., ASCE, Vol. 103, No. HY4: 353–365.

Liggett, J.A. and Liu, P.L-F. 1983. The boundary integral equation method for porous media flow. George Allen & Unwin (Publisher) Ltd., London.

Mizumura, K. 1985. Groundwater flow analysis under space-varying and moving recharges. Boundary Elements VII, Proc. of 7th Int. Conf. on BEM, Vol. 1, Sept: 3.43–3.52.

Muskat, M. 1937. The flow of homogeneous fluids through porous media. McGraw-Hill Book Company.

Nakayama, T. and Washizu, K. 1981. The boundary element method applied to the analysis of two-dimensional nonlinear sloshing problems. Int. J. for Num. Methods in Engrg., Vol. 17: 1631–1646.

Smith, R.E. and Hebbert, R.H.B. 1979. A Monte Carlo analysis of the hydrologic effects of spacial variavility of infiltration. Water Res. Res., Vol. 15, No. 2, April: 419–429.

Chapter 8

Unconfined Groundwater Flow

by G.P. Lennon

8.1 Introduction

The general equation governing unconfined groundwater flow is a transient, three-dimensional partial differential equation with variable coefficients. As summarized in this chapter, for many problems the equation can be simplified such that the associated free space Green's function is known. Then the Boundary Element Method (BEM) becomes a very efficient technique for solving for the location of the transient free surface.

After the formulation is introduced, a number applied problems are presented, including recharge problems, well problems and lagoon seepage problems, all involving unconfined (free surface) flow with a moving, nonlinear free surface.

Unconfined flow problems can also be analyzed by vertical integration of the governing equations. This approximation involves nonlinearities that require area integration or further approximations but eliminates the free surface (see Lafe 1981; Ligget and Liu 1983).

A number of investigators have applied the BEM to a wide range of groundwater problems, including Banerjee and Butterfield (1977), Butterfield and Tomlin (1972), Lafe (1981), Lennon (1980), Lennon et al. (1980a), Lennon et al. (1980b), and Liggett and Liu (1983).

The notation used in this chapter conforms to notation commonly used in the groundwater flow literature.

8.2 Theoretical Formulation

8.2.1 Governing Equations in Domain

Fluid flow in an unconfined aquifer is a complicated phenomenon involving microscopic details to follow actual flow patterns. Assuming that the aquifer is a continuum allows a microscopic Darcy (specific) flux, q, to be defined as the flow rate per unit total cross-sectional area. Then the conservation of fluid mass for an infinitesimal control volume can be written as

$$-n\rho\frac{\partial S}{\partial t} - \rho S(n\beta + \alpha)\frac{\partial p}{\partial t} = \nabla\cdot(\rho\boldsymbol{q}) \tag{1}$$

where n is the porosity (ratio of void space to total volume), S is the degree of

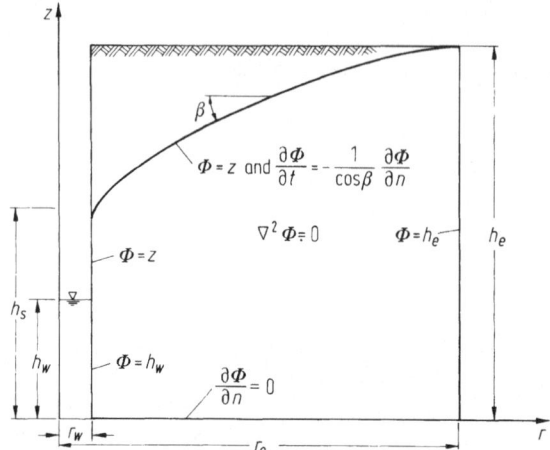

Fig. 1. Definition sketch for well problem

saturation (fraction of voids filled with water), t is time, β is the compressibility of water (inverse bulk modulus of fluid elasticity), and α is the vertical compressibility of the granular skeleton (Eagleson 1970).

For many flow problems in unconfined aquifers, the saturated-unsaturated effects and capillary fringe can be ignored and only the saturated region is considered. The unsaturated zone is separated from the saturated zone by the water table (phreatic surface) along which the pressure is atmospheric, i.e. $p = 0$ (see Fig. 1). When saturated flow is considered ($S = 1$, $\partial S/\partial t = 0$), the first term on the left hand side of Eq. (1) is zero.

The Darcy flux is given by

$$q = -K\nabla\phi \tag{2}$$

where K is the hydraulic conductivity of the porous medium and ϕ is the potential (hydraulic head). Most of the applications presented here assume K is constant.

Anisotropic problems are treated using a scaling of coordinates. Liggett and Liu (1983) discuss the application of the BEM to more general anisotropic, inhomogenous problems in both confined and unconfined aquifers.

The hydraulic head, ϕ, is the total energy per unit weight,

$$\phi = p/\rho g + z \tag{3}$$

where p is the fluid pressure ρ is the fluid density, g is gravitational acceleration, and z is the Cartesian coordinate taken vertically upward. The kinetic energy per unit weight term is small for most applications and is not included in Eq. (3).

Unlike confined flow problems where large pressures can exist, for most unconfined problems the compressibility of the medium and fluid are relatively insignificant compared to the changes in fluid volume resulting from free surface elevation changes. Then we can assume α, β, and $|\nabla\rho|$ are small. Combining Eqs. (1) and (2) for an isotropic homogenous porous medium results in

$$\nabla^2\phi = 0. \tag{4}$$

The storage effect of unconfined aquifers is a result of gravity drainage of water from voids as the water table drops. The effective porosity, n_e, is the fraction of voids that actually drain under gravity, i.e. the porosity minus the moisture retained by effects such as surface tension. The effective porosity enters the formulation in the free surface boundary condition discussed next.

8.2.2 Free Surface Boundary Condition

The free surface elevation is given by $z = \eta(x, y, t)$. From the requirement that $p = 0$, Eq. (3) requires that

$$\phi = \eta \quad \text{on } z = \eta. \tag{5}$$

Defining $F(x, y, z, t) = z - \eta = 0$ and noting that $DF/Dt = 0$, then

$$\frac{DF}{Dt} = \frac{\partial F}{\partial t} + \frac{q \cdot \nabla F}{n_e} = 0 \tag{6}$$

or for isotropic porous media, and noting that $\phi = \eta$ on the free surface,

$$\frac{\partial\phi}{\partial t} = \frac{K}{n_e}\left\{\frac{\partial\phi}{\partial x}\frac{\partial\eta}{\partial x} + \frac{\partial\phi}{\partial y}\frac{\partial\eta}{\partial y} - \frac{\partial\phi}{\partial z}\right\}. \tag{7}$$

Defining the unit outward normal on the free surface as

$$n = \frac{\nabla(z - \eta)}{|\nabla(z - \eta)|} = \frac{\dfrac{\partial\eta}{\partial x}i - \dfrac{\partial\eta}{\partial y}j + k}{\left\{1 + \left(\dfrac{\partial\eta}{\partial x}\right)^2 + \left(\dfrac{\partial\eta}{\partial y}\right)^2\right\}^{1/2}} \tag{8}$$

noting that

$$\cos\beta = \left\{1 + \left(\frac{\partial\eta}{\partial x}\right)\left(\frac{\partial\phi}{\partial x}\right) + \left(\frac{\partial\eta}{\partial y}\right)\left(\frac{\partial\phi}{\partial y}\right) - \frac{\partial\phi}{\partial z}\right\}$$

then

$$\frac{\partial\phi}{\partial n} = \nabla\phi \cdot n = \cos\beta\left\{\left(\frac{\partial\eta}{\partial x}\right)\left(\frac{\partial\phi}{\partial x}\right) + \left(\frac{\partial\eta}{\partial y}\right)\left(\frac{\partial\phi}{\partial y}\right) - \frac{\partial\phi}{\partial z}\right\}. \tag{9}$$

Combining Eqs. (8) and (9) results in a nonlinear boundary condition.

$$\frac{\partial\phi}{\partial t} = -\frac{K}{n_e}\frac{1}{\cos\beta}\frac{\partial\phi}{\partial n}. \tag{10}$$

A dimensionless time is introduced,

$$T = tK/n_e L \tag{11}$$

where L is a characteristic length also used to nondimensionalize the potential,

$\Phi = \phi/L$. For cases where a vertical recharge exists, an extra term occurs,

$$\frac{\partial \Phi}{\partial T} = -\frac{1}{\cos \beta} \frac{\partial \Phi}{\partial n} + W \quad \text{on } z = \eta \tag{12}$$

where W is the recharge rate nondimensionalized by (K/n_e).

8.2.3 Other Boundary Conditions

Two other basic boundary conditions are presented here, the Dirichlet condition

$$\Phi = \Phi_b \quad \text{on } \Gamma_1 \tag{13}$$

where Φ_b is the prescribed potential along Γ_1, a portion of the boundary Γ, and the Neumann condition,

$$\frac{\partial \Phi}{\partial n} = -q_b \quad \text{on } \Gamma_2 \tag{14}$$

where q_b is a specified "flux".

8.3 Numerical Formulation

The major advantage of the direct formulation is that all variables have direct physical significance, i.e. Φ is the potential (hydraulic head) and the $\partial \Phi/\partial n$ is the Darcy flux divided by the hydraulic conductivity. Also, the direct formulation, as used here, requires a single large coefficient matrix associated with the unknowns.

The problems considered here are transient, unconfined two-dimensional, axisymmetric, and three-dimensional problems. The boundary conditions include nonlinearities, sharp corners, and physical flow singularities. As a result, proofs of existance and uniqueness are difficult to obtain. For this reason, the BEM solutions are always compared to available analytical and numerical solutions where possible.

8.3.1 The Integral Equation in Three Dimensions

The divergence theorem states that the divergence of a vector V in a domain D is balanced by the flux associated with that vector through the surface, Γ

$$\int_D (\nabla \cdot V) dV = \int_\Gamma V \cdot n \, dA \tag{15}$$

where n is the unit outward normal of the surface Γ. Using Eq. (15) with $V = \Phi \nabla g$ and subtracting from it Eq. (15) with $V = g \nabla \Phi$, Green's second identity can be written as

$$\int_D (\Phi \nabla^2 g - g \nabla^2 \Phi) dV = \int_\Gamma \left(\Phi \frac{\partial g}{\partial n} - g \frac{\partial \Phi}{\partial n} \right) dA \tag{16}$$

where g and Φ are twice continuously differentiable in D and have continuous first derivatives on Γ (Courant and Hilbert, 1953). Here g is the three-dimensional

Green's function for an infinite space which satisfies

$$\nabla^2 g = -4\pi\delta(x - x_i)\delta(y - y_i)\delta(z - z_i) \tag{17}$$

where $\delta(\ )$ is the Dirac delta function and

$$g = \frac{1}{R}, \quad R = [(x - x_i)^2 + (y - y_i)^2 + (z - z_i)^2]^{1/2} \tag{18}$$

which is singular at $P_i(x_i, y_i, z_i)$. Then Eq. (16) becomes

$$-\alpha_i \Phi_i = \int_\Gamma \left(\Phi \frac{\partial g}{\partial n} - g \frac{\partial \Phi}{\partial n} \right) dA \tag{19}$$

where α_i is 4π times the solid interior angle at a point P_i on the boundary. If P_i is inside the domain, $\alpha_i = 4\pi$.

If Φ and $\partial\Phi/\partial n$ were known everywhere on the boundary, Eq. (19) would lead directly to the solution Φ_i at P_i. For a well posed problem, Φ, $\partial\Phi/\partial n$, or a relationship between them is prescribed (but both are not known). The BEM is used to solve for the missing data for three-dimensional and axisymmetric problems as described next.

8.3.3 Three-Dimensional Problems

A more detailed description of the following procedure is presented by Lennon (1980).

The boundary is discretized into finite elements using an isoparametric formulation where the boundary values and surface geometry are polynomials of equal order. For a linear variation, the global coordinates in a 3-noded linear triangular element are

$$x = \{N\}^T \{x\}_e, \quad y = \{N\}^T \{y\}_e, \quad z = \{N\}^T \{z\}_e \tag{20}$$

where the element shape functions $\{N\}$ are given by Zienkiewicz (1971) as

$$\{N\} = \left\{ \begin{array}{c} \xi \\ \eta \\ 1 - \xi - \eta \end{array} \right\} \tag{21}$$

and the (ξ, η, ζ) coordinate system is shown in Fig. 2. The ζ coordinate is always in the normal direction and does not enter the problem. Also

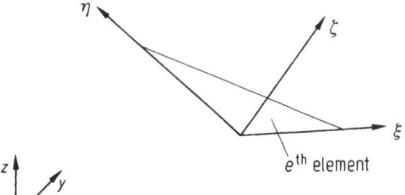

Fig. 2. Relationship between the local and global coordinate systems

$$\Phi = \{N\}^T \{\Phi\}_e, \quad \frac{\partial \Phi}{\partial n} = \{N\}^T \left\{\frac{\partial \Phi}{\partial n}\right\}_e \tag{22}$$

in each element.

Using Eqs. (20) and (22), Eq. (19) can be written as

$$\alpha_i \Phi_i + \sum_{e=1}^{m} \{a\}_e^T \{\Phi\}_e = \sum_{e=1}^{m} \{b\}_e^T \left\{\frac{\partial \Phi}{\partial n}\right\} \tag{23}$$

where m is the total number of elements and

$$\{a\}_e = 2A_e \int_{A_e} \{N\} \frac{\partial (1/R)}{\partial n} d\xi \, d\eta;$$

$$\{b\}_e = 2A_e \int_{A_e} \{N\} \left(\frac{1}{R}\right) d\xi \, d\eta. \tag{24}$$

The numerical evaluation of the integrals in Eq. (24) are performed by Gaussian quadrature (Cowper 1973; Abramowitz and Stegun 1974).

Because the summations involve Φ and $\partial \Phi / \partial n$ at each boundary node, these values must be determined before Φ_i is obtained. To obtain the missing boundary values, Eq. (23) is used N times, with point P_i chosen at each of the N boundary points. When P_i is chosen on the boundary, care must be taken when evaluating $\{b\}$ and α_i as discussed next.

When the point P_i coincides with the third node of an element, the vector $\{a\}_e$ given by Eq. (24) is zero due to the fact that $\partial R / \partial n = 0$. Also b_1 and b_2 contain non-singular integrands and are evaluated using Eq. (30). The integral b_3 is broken into two parts, $b_3 = b_3' + b_3''$. The non-singular part given by

$$b_3' = 2A_e \int_{A_e} (N_3 - 1) \frac{1}{R} d\xi \, d\eta \tag{25}$$

is evaluated using Gaussian quadrative. The part with an integrable singularity,

$$b_3'' = 2A_e \int_{A_e} \frac{1}{R} d\xi \, d\eta \tag{26}$$

can be integrated directly as presented by Wu (1976) and Lennon (1980). A similar procedure is used when the singular node coincides with the first or second node of the triangle.

The solid angle α_i is calculated from

$$\alpha_i = \sum_{j=1}^{L} \tau_j - (L - 2)\pi \tag{27}$$

where τ_j are the angles of intersection of the L triangular elements meeting at point i.

The transformation of the discretized integral equation, Eq. (23), to the system of simultaneous equations,

$$[A]\{u\} = \{c\} \tag{28}$$

is accomplished by performing the summations in Eq. (23). The unknowns $\{u\}$ are either Φ or $\partial\Phi/\partial n$ at each nodal point. For an element with a specified flux boundary condition, the contribution to c_i is

$$-\{b\}_e^T \{q_b\}_e.$$

Also, Φ is the unknown at each node and $\{a\}$ becomes contributions to A_{ij}.

For Φ specified elements,

$$-\{a\}_e^T \{\Phi\}_e$$

is added to c_i and $\{b\}$ are incorporated into A_{ij}. When elements with Φ prescribed and elements with $\partial\Phi/\partial n$ prescribed border a node, the node is a corner node with one unknown value of $\partial\Phi/\partial n$.

When P_i is on a specified flux boundary, α_i is added to A_{ij}. When P_i is on a Φ specified boundary, $-\alpha_i\Phi_i$ is added to c_i.

To form the complete c_i and the entire i^{th} row of A_{ij}, the summation over all elements $e = 1, 2, \ldots, m$ is performed. The appropriate contributions are made based upon the type of boundary condition that is prescribed. The process is repeated for P_i, $i = 1, 2, \ldots, N$, forming the system of simultaneous equations.

The solution for $\{u\}$ is obtained by Gaussian elimination. The entire distribution of Φ and $\partial\Phi/\partial n$ is then known on the discretized boundary, and Eq. (23) can be used to solve for the potential at any point P_i inside D with $\alpha_i = 4\pi$.

8.3.4 Axisymmetric Problems

For axisymmetric problems all physical variables are independent of the θ direction in the cylindrical coordinate system (r, θ, z). Although the Green's function is not axisymmetric, integration can be carried out in the θ direction, resulting in

$$-\alpha_i \phi(r_i, z_i, t) = \oint_\Gamma r \left(\phi \frac{\partial G}{\partial n} - G \frac{\partial \phi}{\partial n} \right) d\xi \tag{29}$$

where G can be considered as the potential generated by a ring source at (r, θ, z), i.e. an axisymmetric Green's function given by

$$G(r, z; r_i, z_i) = \int_0^{2\pi} g \, d\theta = \frac{4}{(a + b)^{1/2}} K(m) \tag{30}$$

(see Jaswon and Symm 1977; Lennon 1980; Liggett and Liu 1983). where

$$a = r_i^2 + r^2 + (z - z_i)^2; \quad b = 2rr_i \tag{31}$$

and

$$K(m) = \int_0^{\pi/2} \frac{d\theta}{(1 - m\sin^2\theta)^{1/2}}; \quad 0 \leqslant m < 1 \tag{32}$$

is the complete elliptic integral of the first kind with modulus m

$$m = \frac{4rr_i}{(r + r_i)^2 + (z - z_i)^2}. \tag{33}$$

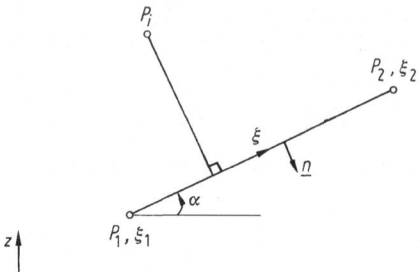

Fig. 3. Definition of the local coordinate system along an axisymmetric boundary element

The discretized form of the integral equation can again be expressed by Eq. (23) except that $\{a\}$ and $\{b\}$ are defined by

$$\{a\} = \int\limits_{\xi_i}^{\xi_{i+1}} \{N\} \left(\frac{\partial G}{\partial n}\right) r \, d\xi; \quad \{b\} = \int\limits_{\xi_i}^{\xi_{i+1}} \{N\} Gr \, d\xi. \tag{34}$$

The ξ coordinate system is shown in Fig. 3 for an axisymmetric segment.

The integrals in Eq. (34) are evaluated using Gaussian quardrature using a polynomial approximation for the elliptic integral (Abramowitz and Stegun 1974). The details of the integration are presented in Lennon et al. (1979a) and Lennon (1980).

The solid interior angle is given by $\alpha = 2\beta$ where β is the interior angle formed by two adjacent line segments.

The solution for the unknowns then proceed in a manner similar to the three-dimensional formulation.

8.3.5 Two-Dimensional Problems

The focus of this chapter is the solution of three dimensional and axisymmetric unconfined flow problems. For details on the two-dimensional formulation, the reader is referred to Liggett and Liu (1983) or Lafe (1981).

8.3.6 Treatment of The Free Surface Boundary Condition

Equation (12) is a nonlinear boundary condition that is applied on the unknown free surface location using finite differences, Eq. (12) becomes

$$\Phi^{k+1} \approx \Phi^k - \frac{\Delta T}{\cos \beta^k}\left[\theta'\left(\frac{\partial \Phi}{\partial n}\right)^{k+1} + (1 - \theta')\left(\frac{\partial \Phi}{\partial n}\right)^k\right]$$
$$+ \Delta T[\theta' W^{k+1} + (1 - \theta')W^k] \tag{35}$$

where the superscript k denotes the time level, ΔT is the difference in nondimensional time between the $(k + 1)^{\text{th}}$ and k^{th} time levels, and θ' is a weighting factor $(0 \leqslant \theta' \leqslant 1)$ positioning the time derivative between the kth and $(k + 1)^{\text{th}}$ time levels.

Equation (35) is applied at the advanced time level $T^{k+1} = T^k + \Delta T$ for all N points. The unknowns are all the $\{\Phi\}$ and $\{\partial \Phi/\partial n\}$ values at time $(k + 1)$ at the new unknown free surface location. By applying Eq. (35), all of the unknown values of Φ^{k+1} are rewritten in terms of the unknown $(\partial\Phi/\partial n)^{k+1}$ leaving N unknown quantities. Collecting coefficients of $(\partial\Phi/\partial n)^{k+1}$ yields as a system of N simultaneous equations with $\{u\} = \{(\partial\Phi/\partial n)^{k+1}\}$ for free surface points. Once the solution for $(\partial\Phi/\partial n)^{k+1}$ is obtained, Φ^{k+1} is calculated from Eq. (35). A predictor-corrector procedure can also be used within a timestep to update the value of $\cos\beta^k$.

8.4 Recharge Problems

Groundwater recharge problems can be analyzed for aquifers of infinite $(h = \infty)$ or finite depths (see Fig. 4). The uniform recharge intensity is denoted by W and the free surface elevation is represented by $z = \eta(r, t)$. The radius of the circular recharge area, L, is chosen as the length scale in Eq. (11). The boundary value problem for Φ can then be described by:

1) Φ satisfies Laplace's equation in the fluid domain.
2) The bottom is impervious $(\partial\Phi/\partial n = 0)$
3) The free surface boundary conditions are given by Eqs. (5) and (12).

The boundary value problem can be readily solved by the numerical scheme developed in Sect. 8.3.

The axisymmetric BEM is first applied to a problem for an infinitely deep aquifer (i.e., $h \to \infty$). Starting at $T = 0$, recharge occurs in the nondimensional area $r \leqslant 1$

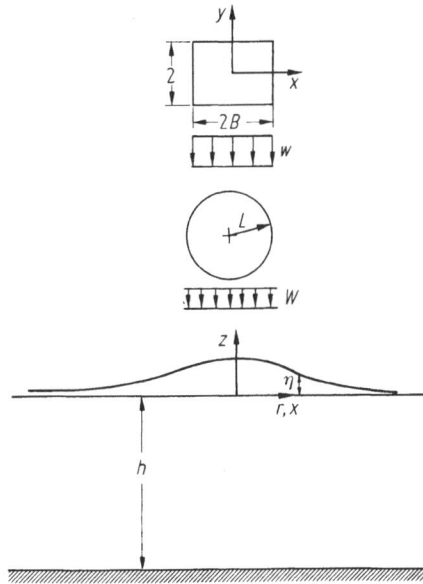

Fig. 4. A circular or a rectangular recharge area causing a rise in the water table

and the water table is at rest at time $T = 0$. Equation (19) is solved at nodes on the entire boundary, including any infinite boundaries. For elements on the infinite boundary, $\{a\}_e^T\{\Phi\}_e \to 0$ because $\Phi \to 0$ faster than $\{a\}_e$ increases. The terms $\{b\}_e^T\{\partial\Phi/\partial n\}$ are also zero since $\partial\Phi/\partial n$ is zero on the infinite boundary. The free surface is assumed to be undisturbed at $r \geqslant r_n$ (i.e., $\eta \approx \Phi \approx 0, r_n < r < \infty$). To insure the truncation of the free surface boundary at r_n has no effect on the solution, r_n was continually moved outward until no change in the solution was detected.

A linearized solution in the far field can be used to approximate Φ and $\partial\Phi/\partial n$ for $r > r_n$ in a manner analogous to the two-dimensional case studied by Liggett and Liu (1983). However, because the axisymmetric problem is confined to a limited area, this increased far-field accuracy is not required.

The BEM results using 22 nodes on the free surface were compared with the linearized theory of Dagan (1967a). Excellent agreement (less than 1%) was achieved for $W \leqslant 0.02$. For a relatively large recharge rate of $W = 0.2$ excellent agreement between the linearized solution and the BEM is obtained for early times; however, at larger times the linearized theory underestimates the peak free surface by about 5% (see Fig. 5).

Calculations are also performed for finite aquifer thicknesses. Figure 6 shows the time history of the maximum water table at $r = 0$ for various values of aquifer thickness and $W = 0.02$. The numerical results agree quite well with the linearized theory. For $h > 10$ the aquifer behaves virtually as an infinite aquifer. For higher recharge rates the BEM results indicate that the linearized theory over-predicts the water table location.

Rectangular recharge regions can be analyzed using the three-dimensional BEM. The normalized peak water table deflection at $(x, y) = (0, 0)$ is shown in Fig. 7

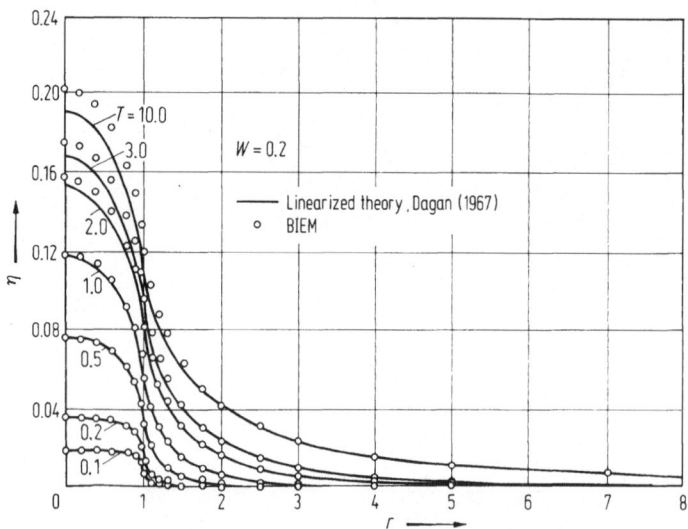

Fig. 5. The phreatic surface for a recharge of $W = 0.2$

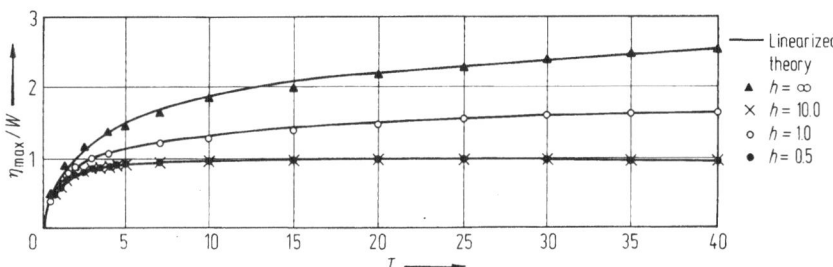

Fig. 6. The normalized maximum free surface rise for various values of aquifer thickness ($W = 0.02$)

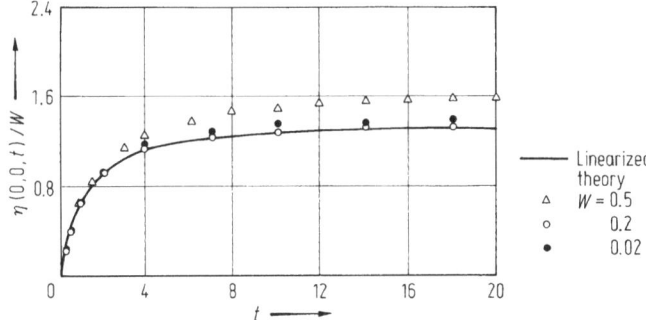

Fig. 7. The normalized maximum free surface rise for various values of the recharge, infinite aquifer thickness, $B = 1.5$

for an infinite depth aquifer, various values of W and a rectangular recharge region with one length 50% longer than the other (see Fig. 4). Lennon (1980) presents many details of the problem including information on the accuracy of the numerical quadrature, the description of the 165 node grid and results for other geometries.

8.5 Well Problems

Radial flow towards a well is characterized by the presence of a seepage face and a vertical boundary very close to the axis of symmetry as shown in Fig. 1. The water table is maintained at a constant level a finite distance away from the well at $r = r_e$.

The boundary value problem for free surface well flows can be described as follows (see Fig. 1):

1) Φ satisfies $\nabla^2 \Phi = 0$ in the flow region
2) $\Phi = h_w$ on $r = r_w, 0 \leqslant z \leqslant h_w$
3) $\Phi = z$ on $r = r_w, h_w \leqslant z \leqslant h_s$
4) $\Phi = z$ on $r_w \leqslant r \leqslant r_e, z = \eta$
5) The free surface boundary conditions given by Eqs. (5) and (12).

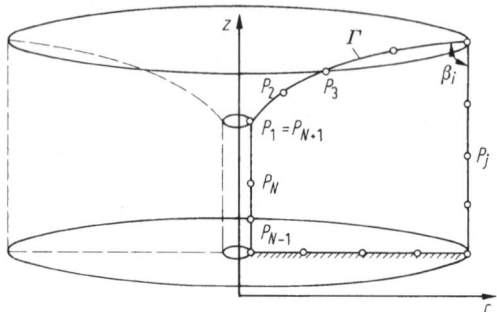

Fig. 8. Axisymmetric BEM grid for the pumping well problem

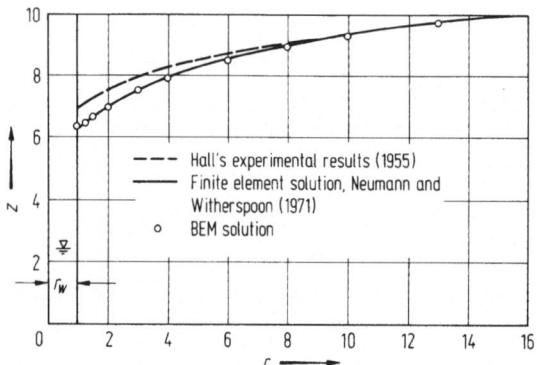

Fig. 9. Comparison of steady-state results (finite aquifer)

A guess at the free surface is made between $r = r_w$ and $r = r_e$ at time T^k. The BEM is used to solve for the unknowns, including Φ^{k+1} by Eq. (35). A new free surface location is then obtained at $\eta^{k+1} = \Phi^{k+1}$. The process is then repreated until $\partial\Phi/\partial n \to 0$ on the free surface, i.e. steady state is achieved.

A sample axisymmetric grid of N points is shown in Fig. 8. The generating contour Γ is rotated through 360 degrees to provide the entire three-dimensional surface. In Fig. 9 the BEM steady-state results are compared with Hall's (1955) experimental results and a finite element solution (Neuman and Witherspoon 1971). The agreement between the BEM and finite element solution is excellent. The elevation of the water table is higher due to capillary effects. For field problems the capillary effect will usually be negligible.

The results shown in Fig. 9 were obtained with 33 nodal points, a well radius $r_w = 1$, and a radial length of $r_e = 50$. Eight Gauss points were used along each element to evaluate the integrals given in Eq. (24). A number of other axisymmetric well problems are reported by Lennon et al. (1979b), Lennon (1980) and Liggett and Liu (1983).

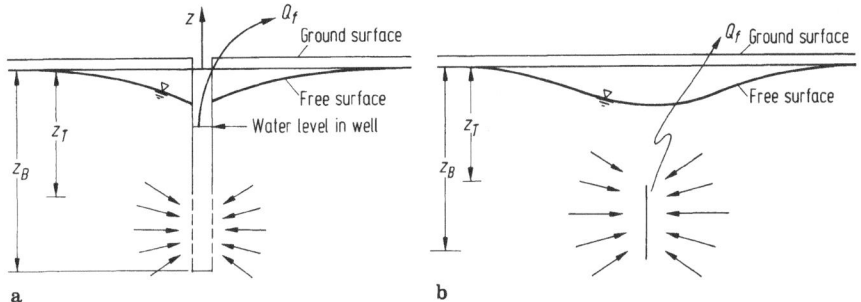

Fig. 10. Diagram of (**a**) a real well and (**b**) a representation of a real well by a line sink

Three-dimensional BEM solutions are obtained by replacing a finite diameter well by a line sink, yielding solutions that are accurate enough for most engineering uses (see Fig. 10). If the solution is required within a few well diameters of the well, the actual well geometry must be modelled as shown in Fig. 1. It is assumed that the vertical well is partially screened between z_B and z_T, the water level in the well never drops below the top of the screen, and the constant flowrate Q_f occurs uniformly across the screen length. The potential due to a line sink of uniform strength $Q_f/(z_T - z_B)$ is further approximated as the potential due to a summation of n_p point sinks using Gaussian quadrature, i.e.,

$$\int_{z_B}^{z_T} \left(\frac{Q_f}{z_T - z_B}\right)\delta(x - x_k)\delta(y - y_k)\delta(z - z')\,dz'$$
$$= \sum_{k=1}^{n_p} \frac{Q_f \omega_k}{2}\delta(x - x_k)\delta(y - y_k)\delta(z - z_k) + E \tag{36}$$

where E is the error introduced using the quadrature (Abramowitz and Stegun 1974). The expression $(Q_f \omega_k/2)$, where ω_k is the Gauss weight, may be interpreted as the strength of the kth point sink at (x_k, y_k, z_k), where z_k is given by

$$z_k = \frac{z_T - z_B}{2}\chi_k + \frac{z_T + z_B}{2} \tag{37}$$

and the Gauss weights are given by $\chi_k, k = 1, 2, \ldots, n_p$.

For n_p point sources or sinks inside the domain D located at (x_k, y_k, z_k) with strengths $Q_k, k = 1, 2, \ldots, n_p$, the resulting integral equation becomes

$$-\alpha_i \Phi_i + \sum_{k=1}^{n_p} Q_k g(x_k, y_k, z_k) = \int \left(\Phi\frac{\partial g}{\partial n} - g\frac{\partial \Phi}{\partial n}\right) dA \tag{38}$$

where

$$g(x_k, y_k, z_k) = \{(x_i - x_k)^2 + (y_i - y_k)^2 + (z_i - z_k)^2\}^{-1/2}. \tag{39}$$

The extra term on the left hand side of Eq. (38) depends only upon the strength of the source and the distance between points k and i. This contribution is added to

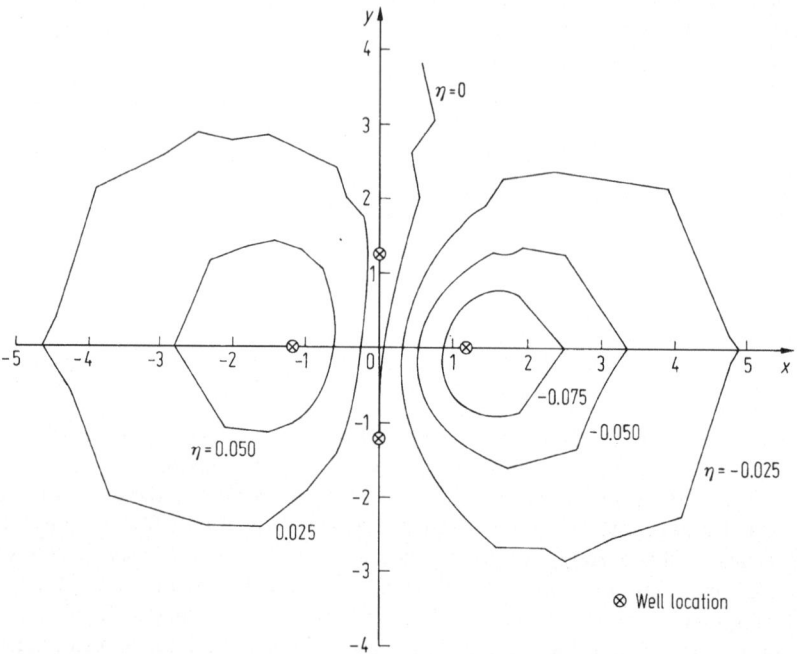

Fig. 12. Steady-state plan view of the phreatic surface solution for the case of four wells of different strengths, aquifer of infinite depth, and spacing of $d = 1.2$

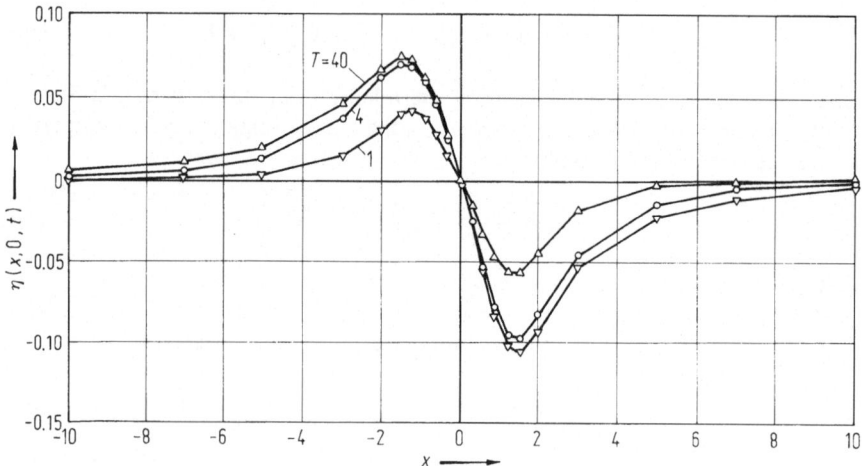

Fig. 13. Phreatic surface solution along the x-axis for the case of four wells of different strengths, aquifer of infinite depth, and spacing of $d = 1.2$

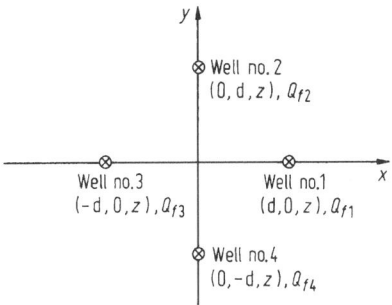

Fig. 11. Plan view of well field; $d = 1.2$

the known right hand side of the simultaneous system of equations. The known surface at the kth time level is used in the numerical calculation of $g(x_k, y_k, z_k)$ so that the governing equations and discretized boundary conditions are linear.

Four wells are located in an infinitely deep, infinitely horizontal aquifer. The well system is basically a discharge/recharge well system with a single well at $(x, y) = (1.2, 0)$ pumping at a rate of $Q_1 = 0.6\,\pi$ and a recharge well operating at 2/3 the strength of the pumping well located at $(-1.2, 0)$ (see Fig. 11). The well at $(0, 1.2)$ is a weakly recharging well operating at 1/4 the strength of the pumping well, and well at $(0, -1.2)$ is a recharge well operating at 1/12 the strength of the pumping well. The free surface contour lines in plan view in Fig. 12 show that the steady-state solution ($t > 80$, $h = \infty$) is basically a two well solution, with a greater free surface deflection at the pumping well than at the recharge well. The line of $\eta = 0$ has swung clockwise from the y-axis because the recharge well is weaker then the dischrage well. The coarse solution for $x > 2$, $y > 2$, is due to the coarse grid used and can be made smoother by using a finer grid. Figure 13 shows that distribution of the free surface elevation along the x-axis for selected times.

8.6 Seepage from a Lagoon

The next problem considered herein is the steady-state flow from a lagoon through a homogeneous porous material down to a water table located at $z = 0$. The condition at the lagoon boundary is $\Phi = $ constant, and the unknown on the lagoon boundary is $\partial \Phi / \partial n$.

The three-dimensional BEM approach is employed in a similar manner as he recharge problems. The boundaries discretized into linear triangular elements. The grid is similar to the recharge grid except the z coordinates are assigned to form the lagoon and an initial guess is made for the sloping free surface. Then the location of the free surface is computed at successive time levels until a steady-state is reached; that is, until successive solutions for the free surface and $\partial \Phi / \partial n$ and smaller than some error criteria. The error criteria used here for solutions at successive time levels with $\Delta T = 5$ are: (1) The z coordinate of points on the free surface, $z = \eta$, change less than 0.1% of z_{WT}, the distance between the lagoon surface and the water table,

(2) the changes in $\partial\Phi/\partial n$ on the lagoon boundary are less than 0.01% of any $\partial\Phi/\partial n$ value, and (3) the $\partial\Phi/\partial n$ on the free surface is less than 0.1% of a typical $\partial\Phi/\partial n$ value through the lagoon surface.

The three-dimensional steady-state free surface solution for $y = 0$, $x > 0$ is shown in Fig. 14 for a water table 4 units below the lagoon surface ($z_{WT} = 4$). The shape of the lagoon is an inverted pyramid of n_s sides, where n_s is taken to be 4, 8, and 16. As n_s becomes large the three-dimensional lagoon solution should approach that of the axisymmetric case (a cone shaped lagoon). Results for the axisymmetric lagoon are also shown in Fig. 14. The results indicate that $n_s = 8$ is quite sufficient to represent the axisymmetric solution.

The flux through the lagoon, represented by a plot of $\partial\Phi/\partial n$, is shown in Fig. 15 for the axisymmetric solution with 12 and 24 nodal points, as well as the three-

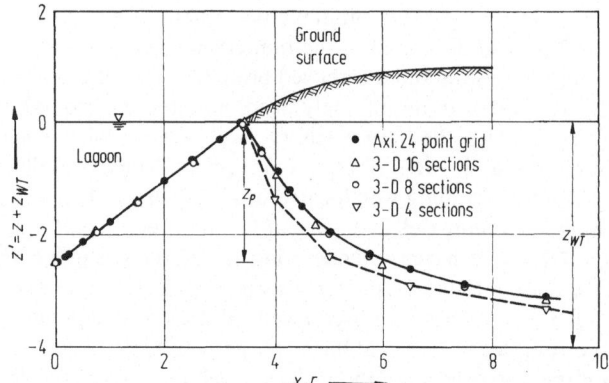

Fig. 14. The steady-state flow from a lagoon down to a water table, $z_{WT} = 4$

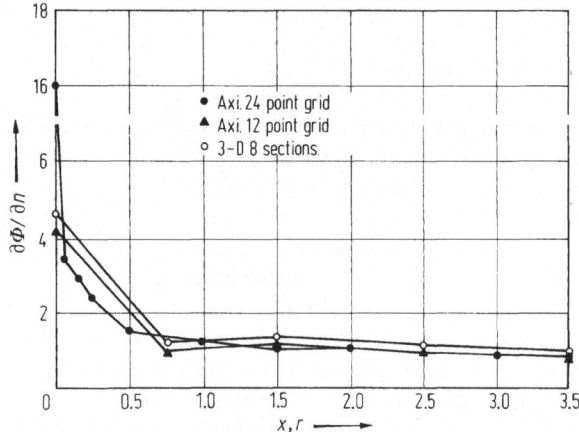

Fig. 15. The comparison of three-dimensional and axisymmetric BEM $\partial\Phi/\partial n$ solutions for the flux through a lagoon

Table 1. Grid type and flow rates, lagoon problems

Grid	N	n^*	Q_f
Axisymmetric	12	—	46.9
Axisymmetric	24	—	47.1
3-D 4 Sections	117	30	38.0
3-D 8 Sections	209	27	44.4
3-D 16 Sections	305	20	47.3

dimensional solution for $n_s = 8$ along $y = 0$ (an edge of the pyramid). The 12 point axisymmetric grid has the same nodal locations as the three-dimensional grid along the x-axis. The three-dimensional solution is slightly higher than the axisymmetric solution along the x-axis. However, along a flat side of the pyramid the three-dimensional $\partial\Phi/\partial n$ solution is, in general, slightly below the axisymmetric solution, yielding approximately the same flow rate, Q_f (see Table 1).

The singular behavior at the tip of the lagoon is better represented by concentrating more elements close to the tip. This effect is demonstrated in Fig. 15 by comparing the sharper axisymmetric $\partial\Phi/\partial n$ profile and 24 points (with a point at $r = 0.05$) to the 12 point axisymmetric grid and three-dimensional grid which have points at $r = 0.75$ and $(x, y) = (0.75, 0)$, respectively.

Between 4 and 8 iterations were required for the solution to reach essentially steady conditions, depending on the initial guess, size of grid, and time step size. As in the three-dimensional recharge problems n^* is the number of unique unknowns using double symmetry to advice the number of unknowns.

Acknowledgements

I gratefully acknowledge the American Geophysical Union, who gave permission to redraw figures originally presented in the following *Water Resources Research* papers:

Vol. 15, No. 5, 1102–1106, Oct. 1979 (Lennon, Liu and Liggett)
Vol. 15, No. 5, 1107–1115, Oct. 1979 (Lennon, Liu and Liggett)
Vol. 16, No. 4, 651–658, Aug. 1980 (Lennon, Liu and Liggett)

References

Abramowitz, M., and Stegun, I.A., *Handbook of Mathematical Functions*, Dover Publications, Inc., New York, NY, (1974).

Banerjee and Butterfield, R., "Boundary Element Methods in Geomechanics," Chap. 16 in *Finite Elements in Geomechanics*, ed. G. Gudehus, John Wiley & Sons, New York, (1977).

Bear, J., *Dynamics of Fluids in Porous Media*, Elsevier, New York, NY, (1972).

Brebbia, C.A., and Walker, S., *Boundary Element Techniques in Engineering*, Butterworth, Inc., Woburn, MA, (1980).

Butterfield, R. and Tomlin, G.R., "Integral Techniques for Solving Zoned Anisotropic Continuum Problems", Variational Methods in Engineering, Vol. II, Proc. Int. Conf. at Southampton, England, (1972).

Cowper, G.R., "Gaussian Quadrature Formual for Triangles," *Int. J. Numerical Methods*, Vol. 7, No. 3, 405–408, (1973).

Courant, R. and Hilbert, D., *Method of Mathematical Physics*, Interscience, Vol. II. (1953).

Dagan, G., "Linearized Solutions of Free Surface Groundwater Flow with Uniform Recharge", *J. Geophys. Res.*, Vol. 72, No. 4, 1183–1193, (1967a).

Dagan, G., "A Method of Determining the Permeability and Effective Porosity of Unconfined Aniso-tropic Aquifers", *Water Resources Res.*, Vol. 3, No. 4, 1059–1071, (1967b).

Eagleson, P.S., *Dynamic Hydrology*, McGraw-Hill, New York, NY, (1970).

Freeze, R.A., and Cherry, J.A., *Groundwater*, Prentice Hall, Inc., Englewood Cliffs, NJ, 1979.

Hall, H.P., "An Investigation of Steady Flow Toward a Gravity Well," *LaHouille Blanche*, Vol. 10, No. 8, (1955).

Jaswon, M.A., and Symm, G.T., *Integral Equation Methods in Potential Theory and Elastostatics*, Academic Press, New York, NY, (1977).

Lafe, E.O., "Boundary Integral Solution to Nearly Horizontal Flows in Multiply Zoned Aquifers," thesis presented to Cornell University at Ithaca, NY in partial fulfillment of the requirements for the degree of Doctor of Philosophy, (1981).

Lennon, G.P., "The Boundary Integral Equation Method Applied to Free Surface Flow Problems in Porous Media," thesis presented to Cornell University, at Ithaca, NY, in partial fulfillment of the requirements for the degree of Doctor of Philosophy, (1980).

Lennon, G.P., Liu, P.L.-F., and Liggett, J.A., "Boundary Integral Equation Solution to Axisymmetric Potential Flows: 1. Basic Formulation," *Water Resources Research*, Vol. 15, No. 5, 1102–1106 (1979a).

Lennon, G.P., Liu, P.L.-F., and Liggett, J.A., "Boundary Integral Equation Solution to Axisymmetric Potential Flows: 2, Recharge and Well Problems in Porous Media," *Water Resources Research*, Vol. 15, No. 5, 1107–1115 (1979b).

Lennon, G.P., Liu, P.L.-F., and Liggett, J.A., "Boundary Integral Solutions to Three-Dimensional Unconfined Darcy's Flows," *Water Resources Research*, Vol. 16, No. 4, 651–658 (1980a).

Lennon, G.P., Liu, P.L.-F., and Liggett, J.A., "Boundary Integral Method Applied to Three-Dimensional Unconfined Darcy's Flow," Conference Proceedings, Second International Symposium on Innova-tive Numerical Analysis in Applied Engineering Science, 47–56, (1980b).

Liggett, J.L., and Liu, P.L.-F., *The Boundary Integral Equation Method for Porous Media Flow*, Allen & Unwin, London, U.K., (1983).

Neuman, S.P., and Witherspoon, P.A., "Analysis of Nonsteady Flow with a Free Surface Using the Finite Element Method." *Water Resources Res.*, Vol. 7, No. 3, 611–623 (1971).

Wu, Y.S., "The Boundary Integral Equation Method Using Various Approximation Techniques for Problems Governed by Laplace's Equation", thesis presented to University of Kentucky, Lexington, Ky., in partial fulfillment of the Degree of Master of Science in Engineering Materials, (1976).

Zienkiewicz, O.C., *The Finite Element Method in Engineering Science*, 2nd Ed., McGraw Hill, (1971).

Subject Index